GW00701591

FORGOTTEN GENIUS

PAUL —

ENJOY THE BOOK

— BEST WISHES

[illegible]

Forgotten Genius¶

A Celebration of Past Engineering Craftsmanship —

ARNOLD J. WADE

UNITED WRITERS
Cornwall

UNITED WRITERS PUBLICATIONS LTD
Ailsa, Castle Gate, Penzance, Cornwall.

British Library Cataloguing in Publication Data:
A catalogue record for this book is
available from the British Library.

ISBN 9781852001544

Printed in Great Britain by
United Writers Publications Ltd
Cornwall.

Acknowledgements

My grateful appreciation to those who freely gave advice, assistance, inspiration and practical help: Arthur Astrop; Evelyn Mathewson; Sandra Meacham; Lorna Russell; A.P. Woolrich; Julian Wade.
Finally to my wife Muriel for her support, forbearance and encouragement.

Nature reluctantly yields to man the animal, vegetable and mineral resources. Using these materials it is possible to transform them into objects which enhance our lives. When their usefulness is extinct, nature will retrieve and regenerate them, so continuing the cycle of growth, maturity and decay.

Contents

Introduction

When gazing at a piece of metalwork made some one hundred or so years ago there is a temptation to dismiss it as a piece of rusty junk. If some time is taken to examine it more closely, in the context of the period in which it was made, some questions may arise. What is it? Why was it made? Who designed and made it? How was it made?

This book is an attempt to answer these questions and to acknowledge the craftsmanship which was practised in bygone days. There are some past engineers whose names are well known but there are many designer/craftsmen who have almost been forgotten. They worked quietly and diligently, not seeking praise, and modestly exercised their skills in the application of ergonomic principles, with an appreciation of aesthetic appeal, to produce their work which satisfies a function.

The period under consideration is mainly from 1740 to 1840 and known as the 'Industrial Revolution'. The activities during that time were influenced by previous events and people, and this historical context is significant to subsequent events. This account is a journey along the various threads of history (with divergence at times) quoting examples along the way where solutions have been made to perceived problems. The cases cited represent a pertinent fraction of the contribution made to our development over a period of very many years.

Britain was 'invaded' by the people of continental Europe who brought their skills and knowledge to contribute to the development of the country. Our magnificent cathedrals were modelled on designs in France; the Flemish introduced cloth making; German miners arrived; Huguenot refugees fled from persecution; control of water required Dutch expertise.

All these rich influences undoubtedly contributed to the British economy but the Industrial Revolution was initiated predominately by native craftsmen and engineers. They caused the great changes which took place, especially in mechanical engineering, and laid the foundations for the standard of living which we enjoy today.

The success of commerce lay basically in the dogged determination, skill and perseverance of men with natural genius. They were practical men who invariably came from humble backgrounds with little or no formal education as youngsters, only learning later in life, as the need arose, to support their innate aptitude for invention.

There were times during the Industrial Revolution when an innovative solution was unacceptable due to the prevailing social or economic conditions, only to be rediscovered years later when the problem became a significant hindrance to progress.

Much evidence of the work in past years from the craftsman in metal has been lost due mainly to the natural tendency of the material to rust and decay. More significantly, items have been consigned to the scrap-heap when they have served their useful purpose, to be replaced by the latest fashion – sometimes inferior – in the name of progress. It is strange that whilst something exists we tend to take it for granted and treat it with indifference, yet when it is gone we feel a sense of loss. A parallel can be drawn with steam locomotives, which we remember as being noisy, dirty and cumbersome but have now become objects of admiration.

We owe a debt of gratitude to individuals and dedicated societies who have recognised the value of objects representing a part of our history and have taken care to preserve them for posterity.

1
Foundations

1:1 Inspiration

Engineering is mainly about looking for solutions to problems. The application of basic knowledge together with skills usually satisfies the situation, but it is not always as simple as that and other factors often influence a decision.

These may be grouped under three headings:

1. Analytical – using existing proven data.
2. Empirical – following established methods of doing things.
3. Inspiration – the intuitive ability to recognise a solution.

The inventor or designer/craftsman of the past did not start with a 'blank sheet of paper', there was always something which went before. The analytical aspect was used by applying existing scientific knowledge based on general principles. For instance, an understanding of the behaviour of steam in relation to temperature and heat helped greatly in designing and developing the steam engine.

It was possible for someone having potential, but without the formal training of apprenticeship, to become proficient by watching and talking to experienced craftsmen. The formal apprenticeship was not available to all those with a natural talent to make things. Having the 'know-how' was an invaluable asset and a good reputation could be made, the work become known and commissions then follow.

The names associated with fine craftsmanship were often those who inspired others in the design and make activity. Both Thomas

Chippendale and Thomas Sheraton produced books of patterns and shapes in the eighteenth century for the benefit of cabinet makers. Such works of reference set the standard for design, and the various craftsmen could concentrate on their fine work without worrying about style. The late Georgian period of 1760-1810 contained exquisite examples of fine craftsmanship, and today we can identify objects made then because they conform to the style appertaining at the time.

In the case of the craftsman working in wood or metal and not having access to the accepted designs of the time, he would have to decide somehow on shape and style. By looking for something better than the heavy, plain designs of earlier furniture he could consider architecture as a source for inspiration. The classical designs of columns originating in Greece spread throughout Europe and such shapes were easily adapted to give an aesthetic quality to legs on a table or the supports for a machine.

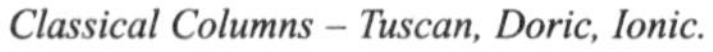

Classical Columns – Tuscan, Doric, Ionic.

Classical influence on the columns of a 19th century beam engine.

Such simulation of classical design also added the dimension of strength, security and social standing to an otherwise utilitarian object. Those fortunate enough to embark on the 'Grand Tour' would return with examples from Italy and Greece of Romanesque, Palladian, Baroque, Rococo, etc. styles, and such ideas were easily passed on to craftsmen to be incorporated in the commissions. Key patterns, which are often attributed to ancient Greece, persist to this day to decorate household objects.

Key patterns – Chinese vessel 1500BC and Roman mosaic floor.

Whilst each period in history had its own identifiable style there were occasions when subsequent generations would copy a style and this created an incongruous impression with that period and became something of an anachronism. This is often shown up when antique experts today quote the true recent age of an 'antique' piece of furniture to a surprised (and disappointed) owner.

Key pattern – Plant pot 20th century.

The inspiration from classical art provides lots of opportunity for decoration, but when faced with a technical problem we must look elsewhere. Invariably, someone somewhere has had the ability to perceive how an unconnected source can provide a 'technological' answer.

Such a source is the natural world, and a survey of how plants and animals behave can have a relevance to engineering. There is, of course, difficulty and frustration when we try to simulate a living organism with man-made materials but solutions are possible by adapting features found in nature in our quest to solve problems.

We cannot (yet) produce a water pump as powerful as an oak tree. Evaporation from the leaves is compensated by drawing water up through the tree from the roots. A fully grown tree will lift 40 gallons up 60 feet during a warm summer's day using only the energy of the sun.

The burdock seed gave rise to the Velcro fastener, but this was only made possible when plastic materials and the associated manufacturing processes were developed. The burdock uses the spines on the seed case to cling to the coat of a passing animal and

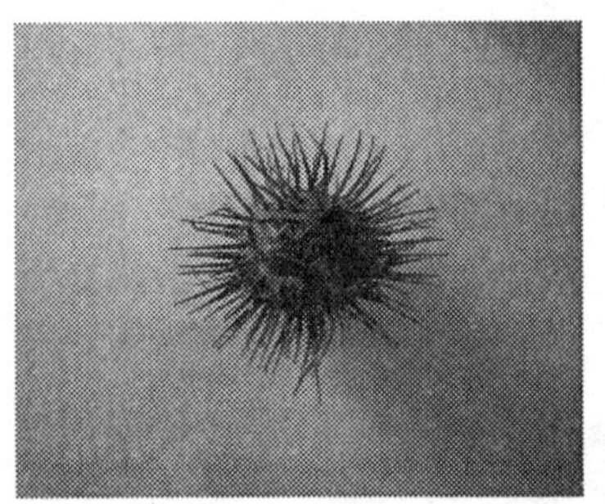

The seed of Burdock with spines magnified

so achieves wide distribution of the seed. Our Velcro fastener is easy to apply, secure in fixing and simple to detach. The German name for the device is 'Klettverschluss' which literally means bur-fastener.

Velcro fastener – hook and loop.

Dissemination of seed by plants using gripping devices to act as a means of transport is very common, and another example is the bur-marigold which employs barbed harpoon-like bristles to attach the seeds tenaciously to the coats of its hosts.

The proportions of the legs of a beetle provide for the heavy body mass to lie below the centre of gravity, this principle was adopted for the underslung chassis of a motor car. A spider's web can, without breaking, stop a bee travelling at 20 miles per hour; but we cannot make such a lightweight material artificially or train the spider to spin such a web commercially! The spider starts the web by spinning radial strands which are used subsequently for moving about. The circular strands are sticky and are used to catch prey. Radial strands have a tensile strength five times that of steel the same size and researchers have recently injected spider genes into the larvae (worm) of the silk moth to give silk greater strength.

The hypodermic needle is found as a wood drill in the wasp Sirex Noctilio. A dragonfly larva can accelerate using the principle of the jet engine. In the crane fly (daddy longlegs) there are knobs on the wings which act like a gyroscope. The propeller blade of an aircraft engine is almost identical to the ash tree seed (key) with a breadth/length ratio of 1:42 and the angle of incidence is the same at all points, i.e. twisted. The main difference lies in the ash key being broader at the end and of thinner material, thus giving a larger working surface with economy of material. We have to use stiffer and thicker material than the ash key as our propeller is a source of power and not a delicate drifting seed.

The honey bee, when building a comb to house its larva, uses a regular hexagon shape for each cell; this structure fully utilises the available space and is very strong. The question arises as to how the bee 'knows' that the honey comb must be made in this way.

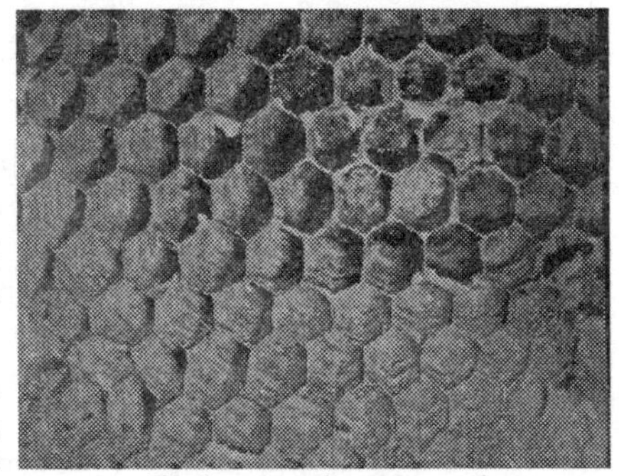

The honeycomb constructed by the honey bee.

The regular hexagon pattern of cells resembles the form taken by a raft of bubbles where the point of contact of the boundaries of three bubbles gives an angle of 120 degrees and this establishes the hexagonal construction. If the body heat of the bee is sufficient to make the wax of the comb soft then the outcome will automatically be identical to the pattern created by bubbles.

Insects which cut leaves use their mandibles in a shearing action to cut, just as we use scissors. The proboscis (tongue) of butterflies and moths is copied, in principle, in the steel tape measure, capable of being tightly coiled but rigid when extended. The ship worm, which did vast damage to the timbers of sailing vessels, avoided being engulfed by the dust it created by having a cowl over its head, and this principle was adopted by Marc Brunel to protect his workers whilst cutting the tunnel under the River Thames in London. Seed pods of the flowers lupin and broom will twist and sharply crack open when ripe under the influence of the hot sun and thus provide the potential energy to create the acceleration to spread the seed widely.

Proboscis of a moth – about to extend to suck nectar.

Many of the examples above have been applied in the past and we still continue to be inspired by nature's achievements. Recent advances in surgery using tiny robotic devices to examine the human body include an extremely fine tube based on the egg-laying probe of a parasitic wasp. The action of the geometer moth caterpillar, known as a looper, using only two pairs of legs to get along, is copied in an exploratory device to crawl along blood vessels during medical examinations.

In addition to the natural world there are other sources for possible inspiration, and one significant area lies in the preparations for war, where a nation is desperate to oppose aggression. Such activity provides an impetus for invention, and subsequently when peace is declared the 'spin-off' can be valuable. A case in point is the silicon chip which transformed lives in the 20th century.

Before leaving this brief comment on sources of inspiration it is important to mention the Golden Ratio. Also known as the Golden Section or Golden Number, this is the ratio 1:1.618 and it is generally accepted that it was first noted by Euclid, although the builders of the Great Pyramid of Giza, in Egypt, incorporated the Golden Ratio in the proportions of the King's Chamber some 2,000 years previously. The Greeks used it in building the Parthenon at Athens. It occurs in the natural world where proportions 'look right'; in trees and flowers and the proportions of our own bodies. The mathematical basis for the ratio is due to Fibonacci [1170-1250], the mathematician known as Leonardo of Pisa. He discovered the series in which each number is the sum of the two which precede it and the ratio of any adjacent numbers closely approximates to 1:1.618, i.e. 1, 2, 3, 5, 8, 13, 21, 34, 55. . . . The ratio, when applied to rectangular objects, gives an aesthetic appeal and such proportions are pleasing to the eye. Examples are to be found in buildings, especially in the shape of window frames, when the ratio of width to height is 1:1.618, otherwise they would appear too wide or too slender.

Having the inspiration for his work, from whatever source, the craftsman had then to create. Being sensitive to the raw materials, the existing methods and techniques were then honed and practised to perfection. We could argue that the craftsman of old used elaborate decoration on the simplest of objects when compared to the trend today of straightforward utility. Those craftsmen took a greater pride in their work, delighting in imparting beauty to a mundane object. Perhaps they had more time to think and consider because they were not constrained by our current demand for instant productivity at minimum cost.

Joseph Whitworth [1803-1887], one of the great engineers of the 19th century, promoted engineering education by endowing colleges and funding scholarships. His axiom was: "The luxuries of one age should be looked upon in the next as the necessary conditions for existence."

The relationship between knowledge, learning and skill can be difficult to understand, and if we combine these with intelligence and ask people to define the differences the answers will vary widely. Put simply, if we ask someone to make or invent something, they will initially have the knowledge or inspiration, then apply skill through learning to achieve a practical result.

The quotation by Alexander Pope that 'a little learning is a dangerous thing' is often mis-quoted as 'a little knowledge is a dangerous thing'. Something which is part learnt, be it a practical skill or a more abstract activity, could be potentially 'dangerous' or at least unsatisfactory. Whereas the acquisition of knowledge, however small, is of value.

1:2 Innovation

It is said that innovative design is 10% inspiration and 90% perspiration. The majority of us when presented with a commission would be hard pressed to draw fully on the knowledge, inspiration, learning and skills to come up with an eminently successful solution. During the late 18th century, a number of engineers had that spark of genius and contributed to the period we now know as the Industrial Revolution.

Prime among these were the inventors of the steam engine, and although each was successful, there were outside influences which frustrated their efforts.

Captain Savory, a military engineer, took out a patent in 1698 according to which, was for a fire engine for 'raising water by the impellent force of fire'. His engine did not work. Thomas Newcomen [1663-1729] then came along but was hindered in his work to produce a steam engine by the wide ranging description in Savory's patent. Any 'fire engine' made would result in royalty payments and even after Savory's death a syndicate continued to receive monies until the patent finally expired in 1733.

Thomas Newcomen was a practical man, an ironmonger, living in Dartmouth, Devon. He was treated with contempt by the scientists of the day and he did not belong to the apparent elitist group of engineers. He was nevertheless able to marry scientific principles with practical engineering to produce his atmospheric engine in 1698. This was not a steam engine as we know it today, since the power was provided by atmospheric pressure and not by the direct action of steam. Newcomen discovered this by accident, when cooling water in the cylinder jacket leaked through a faulty solder joint into the cylinder and caused a vacuum, thereby allowing atmospheric pressure to push the piston down. The cylinder did not have a top and was therefore open to the atmosphere.

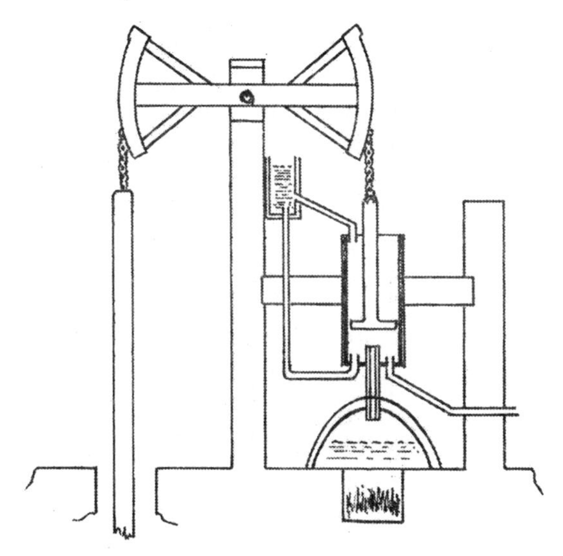

Newcomen's atmospheric engine – early 18th century.

James Watt [1736-1819], in 1765, realised that it was better to condense the steam outside the cylinder and, more significantly, to drive the piston using the steam power. He then set about establishing the renowned company of Boulton & Watt, with centralised manufacture based in Birmingham. Patents were drawn up to protect the steam engine design and the mine owning customers, in addition to purchasing the engines, were obliged to pay one third of the money saved on the cost of the previous fuel used. This was called 'duty' and paid in perpetuity. Despite this, the savings made by the mine owners were significant.

By 1783 the Boulton & Watt engines had completely replaced the Newcomen engines in Cornwall and a number of Cornish engineers then decided that they could make engines themselves. The greatest demand for engines was in Cornwall at that time, where the concentration of copper mines was more substantial than the coal mines elsewhere in the country and all needed engines to pump out water. The Cornishmen all fell foul of infringement of the Watt patent rights until the patent ran out in 1800.

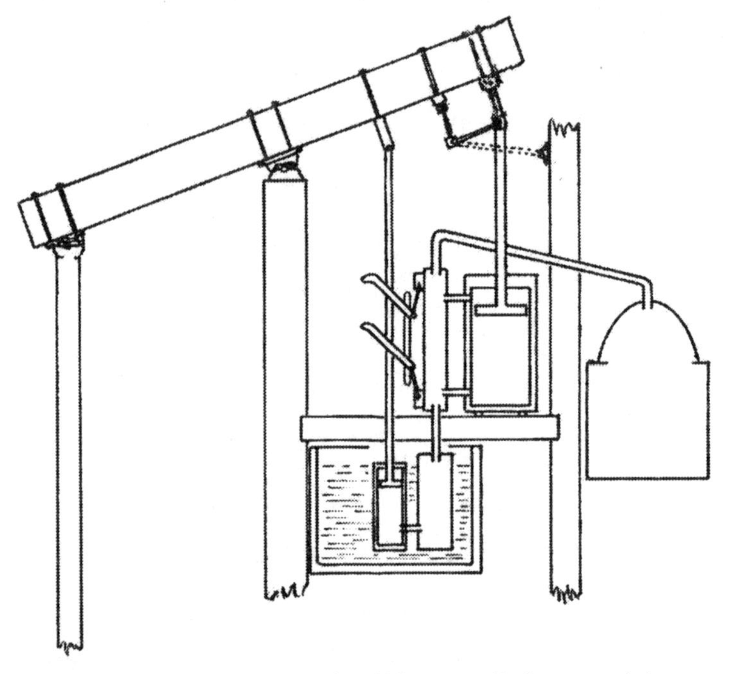

James Watt's beam engine – late 18th century. (Outline principle.)

One man who suffered from the restrictive Watt patent was Richard Trevithick [1771-1833]. His contribution was impressive, using high pressure steam to design and make the first successful steam road carriage and the first steam locomotive. He had the ability to develop new ideas and see them through to fruition but his business strategy was virtually non-existent and very little profit was made. This was in contrast to Boulton & Watt, who pursued a ruthless policy of gain, often going to law to obtain payment and protect their patent rights. Trevithick's name is not as well known as James Watt's who sought publicity and esteem.

Whilst Watt, Trevithick and others were concerned with developing the steam engine, initially to drain or haul in mines and then for wider application, there were engineers who, independently, worked to invent and develop the machines and equipment to make the component parts of the steam engine. It is a fact that the accuracy of tools must be greater than that of the products they produce. This aspect of craftsmanship, known as precision engineering, grew from the needs of the late eighteenth century and the founding father may be regarded as Joseph Bramah. In all, there were eighteen patents attributed to him including the water closet, screw propeller and hydraulic press. His most outstanding achievement was in 1784 when he made a tumbler lock without wheels or wards and offered a reward to anyone who could open it. The lock was finally 'picked' some years after his death.

Lock design was developed by an ironmonger named Jeremiah Chubb, who gained Royal approval and obtained patent protection in1818. Improvements were made by Charles Chubb by adding thief and fire-proof safes to a range of locks and the company grew to gain wide recognition

Most notable amongst Bramah's employees was Henry Maudslay [1771-1831] who started his working life as a carpenter but went on to run a flourishing machine tool business with a string of innovative inventions. In turn Maudslay employed men who would subsequently start up in business on their own account producing machine tools. So, in the early years of the nineteenth century tremendous progress was made by Joseph Clement, Richard Roberts, Joseph Whitworth and James Nasmyth. These men, who made such a valuable contribution to the formation of the machine tool industry, often came from humble backgrounds and were not influenced directly in their chosen profession by their forebears.

One example from this time can serve to show how these early engineers arrived at an apparently simple solution by lateral thinking, which is typical of the genius of the innovator. James Nasmyth [1808-1890] noticed that the conical shaped seating of the safety valve on a steam engine often stuck in the closed position and was unreliable and potentially dangerous. He designed a spherical shaped valve which did not stick. A further advantage was that there was no opportunity for tampering as was the case with the valve used previously which was controlled by a weight on a steelyard type lever.

Much earlier lived one of the greatest innovators of all time, Leonardo da Vinci [1452-1519]. He worked primarily in Italy and, at the end of his life, in France. In addition to his fine work as an artist and his interest in anatomy, he produced engineering ideas some four hundred years before their general acceptance.

All the elements found in the motor car were familiar to Leonardo, e.g. drive devices, joining methods, gearing, pistons and cylinders. It would appear that Leonardo, with his fertile mind, invented mechanisms and produced drawings but did not have them made. This may have been due to the lack of resources at the time, in terms of materials and manufacturing methods, but recent attempts to make some of the ideas using Leonardo's drawings were not always successful.

Many designs are developed and improved over a period of time, matching the parallel progress in scientific experiment, materials availability and the creation of new techniques.

A case in point is the suction pump, used for drawing water, where greater accuracy in manufacture led to increased efficiency and then an understanding of the vacuum relating to air and water pressure. This was complemented by the developments in pure and applied science and so made the foundations of modern power engineering.

Returning for a moment to the late medieval period, it is worth mentioning the achievement of Brunelleschi. He built the dome of the cathedral in Florence, Italy, completed in 1436, without any support work, to a height of sixty feet. The dome is unique because the courses of brickwork lie in a spiral pattern, and the fact that it is still standing some six hundred years later is testimony to the innovative use of brickwork in a self-supporting dome. Some years ago a model of the dome was attempted in order to ascertain the feasibility of the construction method. Unfortunately the model was not successful and clearly the techniques used by Brunelleschi have been lost and not replaced by our advanced modern technology.

There are a number of instances where the skills and knowledge of the past have disappeared and one example is papier-mâché, when in Birmingham during the mid 19th century tables were made with decorative mouldings entirely in this material. The finish was a number of coats of lacquer and the structure was as substantial as wood.

The best examples of decoration in terms of aesthetic appeal are to be found on firearms, where exquisite chasing and tracery on both metal and wood was developed to a fine degree. The locksmith also, like the armourer, created elaborate artwork on locks and associated door furniture. Such embellishment is also found on clocks, watches, saw frames and surgeons' equipment. The stem of a door key would be shaped like a Corinthian column surmounted by classical figures such as griffins placed back to back to form the handle. Escutcheons around key holes were very decorative with natural forms and may be seen today on many cathedral and church doors. Elaborate carving on wooden hand tools was not popular in England where the tool handles just had simple yet graceful lines. This was in contrast to the tools made in Southern Europe, where the Spanish and Italian people had a penchant for decoration and used their long tradition of art with flair at every opportunity.

Fine chasing on a gunstock. 17th century.

Pocket watch balance cock with a 1p coin.

In the Middle Ages the most educated people were the monks, they were able to write when the majority of the population was illiterate. The craftsman, so capable when using his practical skills, usually could not record his work, so very few records of the 'how' and 'why' have come down to us over the years. There were obviously also times, in those far-off days, when a craftsman would be reluctant to share his hard won methods with third parties for fear of competition and then possibly being deprived of his livelihood.

Those who chose to publicise descriptions of their work would have to engage the services of a literate scribe, a scholar and possibly a member of the Church. The difficulty was that the scribe would, no doubt, have little understanding of craftsmanship or how the artefact was made and consequently the record would be inaccurate and incomplete. Unfortunately the craftsman was unable to check the validity of the writing.

There is a parallel between Christian monks, with their erudite pursuits of long ago, and the Ministers of recent times who were able to instigate innovation and develop initiatives for improvement. The clergymen were intellectually equipped to ponder science and draw conclusions about possible practical applications; they were well placed to do this after attending to the spiritual needs of their flock. The Rev William Lee, at Calverton, noticed the technique his wife used whilst knitting and designed and made a stocking frame in 1586. This was not taken up commercially and it was more than two hundred years later, when industry was gathering momentum, that it came to fruition. Another example is the Rev Francis Bashforth of Horncastle, who was also a professor of applied mathematics and invented a chronoscope to measure the speed of a projectile for the Royal Artillery.

Many early industrialists held strong religious views but were often dissenters of the established church, or sometimes they were Quakers; especially Abraham Darby of Coalbrookdale, who pioneered the production of cast iron using coal in the 18th century and ran his most successful business with a strict Quaker background.

Those craftsmen, armed with the necessary knowledge and skills and having been inspired, then had to 'take the initiative' to achieve a result. Initiative is a two-edged sword. If used to change the accepted order, adjust or develop for the better, then it is admired and applauded. On the other hand, if changes are seen as less than acceptable, the outcome will be deplored. Thus initiative can result in success *or* failure.

The great innovators of the late 18th and early 19th century were not merely making replacements for traditional crafts but inventing equipment that was capable of previously unimagined accuracy and performance. They were often working at a time when the world as a whole did not take advantage of their efforts and only later would the benefits be felt.

An inventor required an extremely resolute character, with utmost dedication to the cause, to achieve success. Often external forces from competitors would be sufficient to make the less committed abandon a project were it not for the sure knowledge that the end justified the means.

The pioneers in engineering who, by their ingenuity and persistence, contributed so much to improving our standard of living were:

1755	John Harrison	Portable Timepiece
1770	Ramsden	Sextant
1770	Hargreaves	Spinning Jenny
1771	Arkwright	Cromford Mill
1774	Wilkinson	Boring Machine
1781	Watt	Rotative Steam Engine
1784	Bramah	Secure Lock
1787	Wilkinson	Iron Barge
1795	Bramah	Hydraulic Press
1800	Maudslay	Screw Cutting Lathe
1802	Trevithick	Steam Carriage
1805	Maudslay	Micrometer
1817	Roberts	Planing Machine
1820	Whitney	Milling Machine
1820	Clement	Planing Machine
1830	Bodmer	Gear Cutting Machine
1834	Nasmyth	Shaping Machine
1835	Whitworth	Lathe
1838	Brunel	'Great Western'
1839	Nasmyth	Steam Hammer

There are many more people who may be classed as Inventors/Craftsmen but records of their achievements are either sparse or lost. This list represents those people who were significant in pushing out the frontiers of technology, and the items mentioned represent just a part of their versatility for invention.

2
Processes

2:1 Hand Tools

The basic woodwork tools in ancient times were the axe and the adze. The axe was used for felling, converting and squaring timber and the adze for subsequent surfacing, truing and levelling. In skilled hands the squareness and surface quality of a finished beam was remarkably good. The use of an adze has diminished over the years, the main application being the production of ships' timbers, but during the early 20th century the tool was to be used again. This was by Robert Thompson, known as the Mouseman of Kilburn, who made oak furniture of outstanding quality for churches, cathedrals, schools and stately homes. The adze blade is set at 90 degrees to the handle and is used by the craftsman standing on the work and cutting the wood by swinging the tool between his feet. Robert Thompson made table tops using the adze and at first glance a finished top looks perfectly flat but on closer inspection it is found to consist of a slightly rippled effect. He used the adze as described above with great precision to cut wafer thin shavings 0.015in thick to create an undulating surface finish similar to broadly beaten metal.

Table top cut with an Adze.

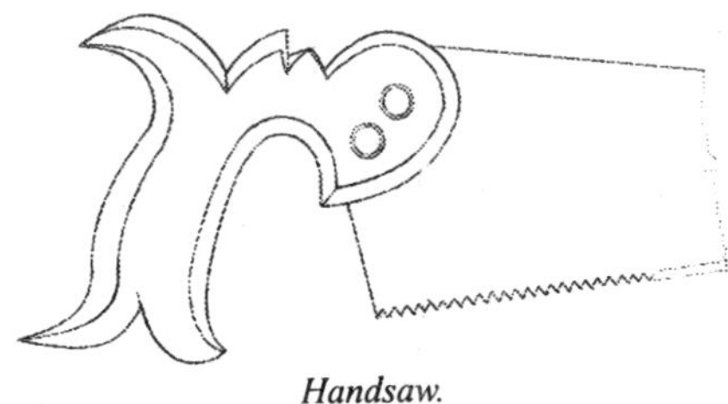

Handsaw.

An alternative to the axe for cutting timber is the handsaw, and early examples with copper blades have been found in Egypt. These early saws were used to cut on the pull stroke to avoid buckling of the blade, and although this technique is less efficient than using the body weight to cut by pushing, it is still used on saws with paper thin blades to achieve delicate cabinet work. The practice of cutting on the push stroke probably came about in Roman times when iron was used to produce stiffer blades. They also set the teeth to project alternately to the side to create a slot – the kerf – which helped to discharge the sawdust and prevent the blade from binding in the wood. Further, they reinforced the back of the blade and also introduced the frame saw which was more adaptable and could cut curves; two styles are evident, both being easy to make with a simple yet effective method to tension the blade.

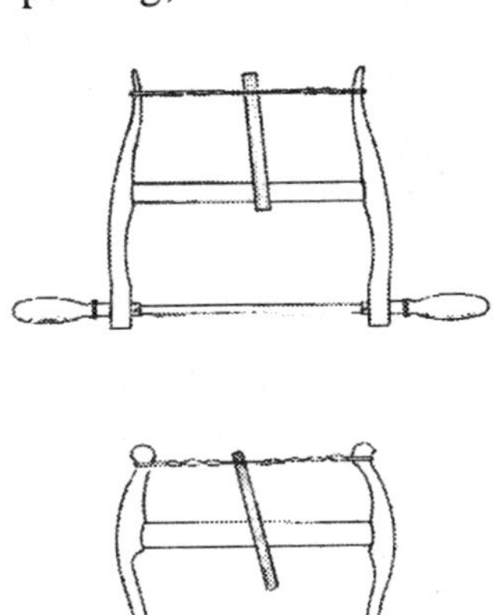

Frame saw – two styles of design.

During the Dark Ages it would seem that the saw disappeared and the axe and adze remained the tools for working wood. The Bayeux Tapestry of 1100AD shows the tools used to make the ships of William the Conqueror's fleet and there is no evidence of a saw. Before mechanisation, the conversion of a tree into usable planks was an arduous job, using animals as well as manpower to manoeuvre a massive mature oak trunk into a position suitable for conversion. The saw used to cut the timber into planks was the aptly named pit saw. This was possibly 12 feet long with a handle at each end and coarse teeth to cut along the grain, also known as a rip saw. In operation the timber trunk was positioned over a pit, and with one man holding the saw above ground his mate would work the other end by standing in the pit. Veneers as thin as one eighth of an inch could be cut from a large tree trunk. The man on the lower end of the saw would be continually showered with sawdust and no doubt found it beneficial to wear a wide brimmed hat!

The teeth of a saw must have the correct degree of hardness, if too soft they would quickly lose their sharp edge and if too hard and brittle they may fracture and also could not be easily sharpened. Saws for cutting wood are not required to be as hard as those used for metalworking and the teeth are easily sharpened using a file which is triangular in profile (to fit the gap between the teeth) and known as a 'three square file'. Woodworkers do not use files to work wood!

On some metal cutting operations a toothless saw is used. This can be a soft mild steel disc a quarter inch wide having a large diameter and rotating at a surface speed of 20,000ft per minute, i.e. more that 200 miles per hour. It is fed into the work at a rate of one inch per second and the cutting works by abrasion and local fusion where the metal being cut is heated by friction to melting point and then rubbed off. Hardened steel is easier to cut than soft steel due to its relatively low melting point and a block of cast iron 8 inches square could be cut through in half a minute.

The handles of wood saws are worth looking at, not only for their pleasing shape but also because they are ergonomically designed. The features add balance and grace to an elegant design inviting the hand to grasp it. The handle aperture sometimes appears too small to accept the four fingers of the craftsman's hand until it is realised that the index finger is placed along the saw back and helps maintain a straight cut. In addition to this the elbow must be placed directly in line with the saw blade and the head held over the saw such that the nose is exactly vertically over the blade. An instructor at college teaching fine woodworking techniques would, on occasion, impress the importance of adopting this posture to the students by saying that if they had a head cold the 'dewdrop' on the end of the nose should be deposited perfectly on the saw back! Presumably such an event would help to lubricate the blade as it cut through the wood.

The circular saw originated in Holland during the 17th century when the ever resourceful Dutch adapted corn processing mills to saw timber, initially with a reciprocating blade. In 1775 Walter Taylor used the circular saw to cut the pulley blocks for the Royal Navy when working for Marc Brunel. This work, incidentally, was a pioneering example of mass production. Taylor then went on to set up his own sawmill and invented a machine to cut wood veneers that became so successful, as an aid to cabinet makers, that the price of furniture was drastically reduced.

The Romans – once again – were apparently the first definite users of the smoothing plane and examples have been found at Pompeii. The modern plane differs in detail but not in principle or general design. The effectiveness of a plane to produce a flat, level surface is 'in built', the carpenter simply ensures that the sharp blade projects very slightly from the sole plate (which is perfectly flat) and then proceeds to cut wafer thin shavings from the wood until the whole area is covered. The conventional plane was adapted with a flexible sole plate to allow both concave and convex shapes to be cut. Also to produce mouldings on the edge of wood with a variety of different shaped blades. The advent of the router made this work much easier and did away with the skill previously required to use the moulding plane.

The business of cutting round holes in wood has seen a long history of all sorts of methods with varying degrees of effectiveness. The auger, originating from the Iron Age was a half pipe, sharpened on the inside edges with a pointed end. This was the traditional boring tool used by shipwrights and wheelwrights and its name derives from the 12th century English NAFU-GAR, a nave piercer (a nave being the hub of a wheel). The letter F has been dropped and the letter N lost by confusion, therefore, a nauger or an auger! The auger developed into a spoon shaped cutting bit with a short wooden cross handle to drive it. In the Middle Ages the breast auger appeared, so now the carpenter could exert his whole body weight to apply pressure to the cut. The auger tended to clog with shavings on deep holes and eventually helical flutes were added to automatically eject the shavings. There are a number of different designs of drills for wood, including spade bits and forstner bits and also the hole saw.

The modern twist drill, primarily designed to cut metal, has helical flutes to assist removal of waste material (swarf) with two cutting edges and dates from 1860.

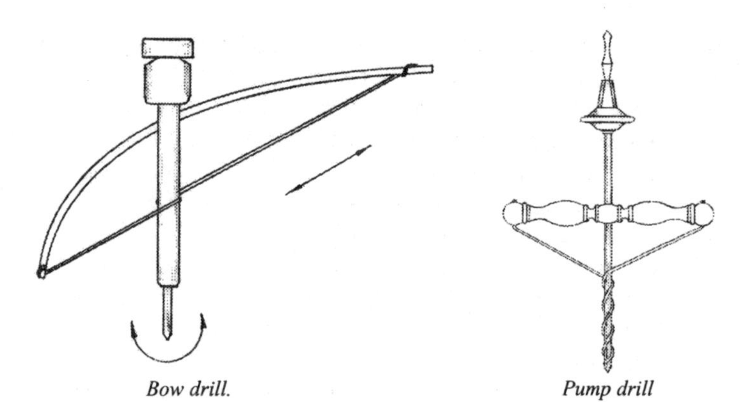

Bow drill. *Pump drill*

The bow drill has a long history and is capable of high speed, this being particularly useful in drilling small holes in hard materials like porcelain. A useful additional feature was its ability to create fire by generating high friction and it is still used for this purpose in some parts of the world today. Similarly, the pump drill is of ancient origin and has a flywheel to store energy and maintain the high speed of rotation.

The earliest record of a brace and bit is said to appear in a triptych by a Flemish artist in 1425. The brace is a crank, changing reciprocal motion – the carpenter's arm moving backwards and forwards – into rotary motion with the bit turning around.

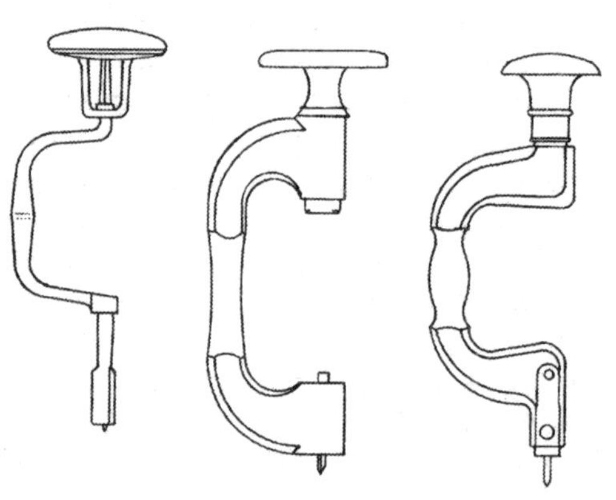

Selection of carpenters' hand braces.

Carpenters' braces varied in design dependent upon the trade for which they were used and most tended to have similarities in the materials of which they were made. A combination of wood and brass gives a nice appeal to the tool, but the real reason for their use is that the metal helps the durability and gives strength and support where the wood has a short grain and would otherwise tend to split.

Most hand tools for cutting became mechanised to reduce the

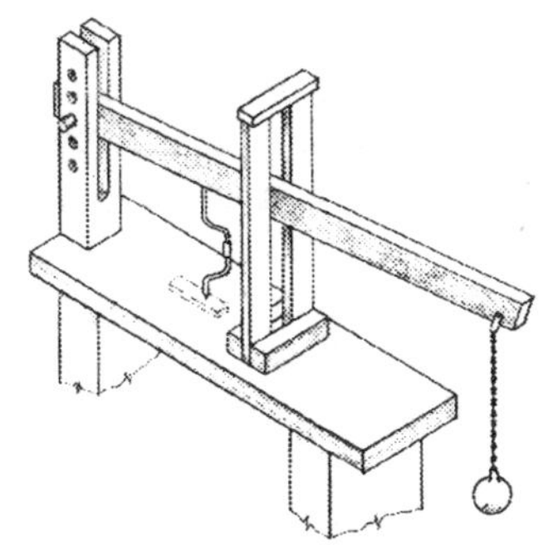
Smith's beam drill.

human effort required to operate them but also served to diminish the skill. One early example is the smith's beam drill, where presumably the blacksmith found it easier to drill a hole in iron rather than to use his traditional method of piercing under heat. This particular drill was the forerunner of the radial drilling machine.

The first mallets were shaped like skittles and had a short life due to the constant pounding on the side grain of the wood. Eventually, of course, a handle was fitted to allow the blows to be taken on the end grain and so prolong its life. The shipwright's caulking hammer is of an interesting design in that it has a split in

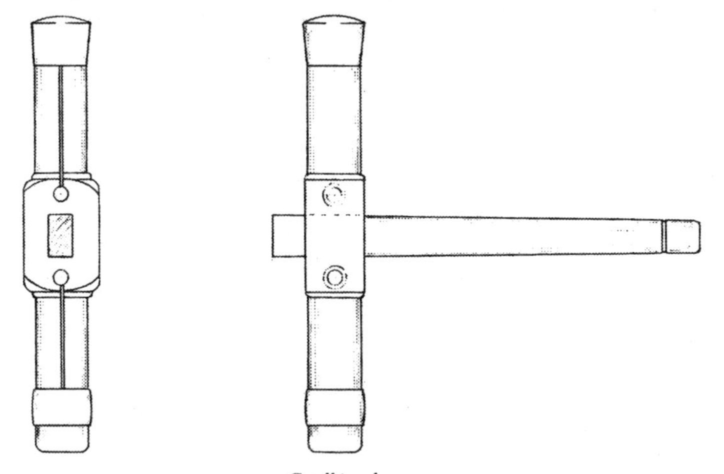
Caulking hammer.

the head. The purpose of this is to deaden the sound and provide a spring to absorb shock and prevent injury to the hand, arm or spine of the shipwright. One can imagine the deafening noise made by a number of men when caulking the deck of a large sailing vessel using conventional hammers. Hammer handles in general are made of a wood which is strong yet resilient enough to absorb the shock of the blow. Too hard a wood, like oak, would easily fracture and more appropriate species are ash and sweet chestnut.

The timber beams in older buildings are joined usually with a mortise and tenon joint, where a projection on the end of one beam (the tenon) enters a matching rectangular socket (the mortise) on the mating beam. No glue or metal fixing was used, so to keep the beams together round dowels were driven through the joint and these are evident as projections on the faces of the beams. In order to make these dowels, in the absence of a lathe, the craftsman would prepare them by hand or to speed things up, use a nogg or moot. This was a simple device; a block of wood with a hole into which a blade projected and the rough wood for the dowel was simply pushed through. Any imperfections of roundness would be smoothed out when fitted into the hole. The dowels were coated with linseed oil prior to being driven in; maximum tension was achieved when they squeaked.

The draw knife is an ancient tool used particularly by wheelwrights to shape spokes for cartwheels. The modern equivalent is the spoke shave. In use the work is held on a

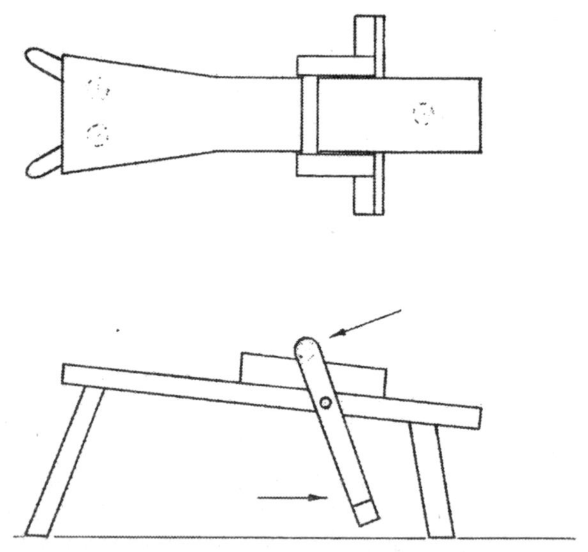

Drawbench – feet push lever which then clamps the work.

drawbench which the craftsman sits astride, pushing with his feet on the bottom of the lever which clamps the work, and makes the cut by drawing the knife towards him. The knife has a handle at either end, at 90 degrees to the blade. The clever aspect of the design of the bench is that the heavier the cut by the knife the greater is the opposing force by the user's feet to grip the work; matching the two forces, pushing and pulling. This technique of using opposing forces is also used in the old design for a wrench where a tapered wedge grips more tightly as greater force is applied to turn the wrench.

Another use for the draw knife, apart from making timber round, was in stop chamfering. This is the process of breaking the sharp edge at an angle of 45 degrees on a length of wood but stopping a short way from the end. An example of this is best appreciated on a cruciform shape where the chamfer is relatively small in comparison with the overall object but the effect is seen and creates an impression reminiscent of the ancient Celtic crosses. When the stop chamfering was applied to a wagon's sideboards it had the effect of making the wood look lighter and also splinters were less likely to occur as hands were run along the board. By stopping a chamfer short of the end of a beam where it enters a joint any weakness is avoided.

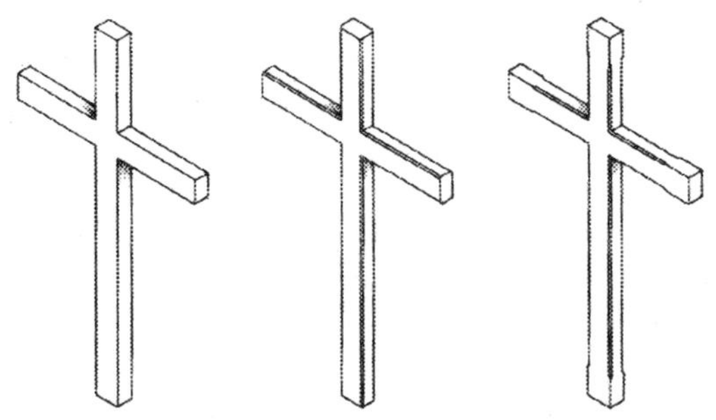
Stop chamfering – showing improvement in appearance.

Craftsmen usually made their own handles for tools, especially turning tools for use on the lathe. They were styled specifically to be comfortable in the hands of the user, sometimes textured for 'grip' and shaped with sensitivity to the run of the wood grain. They became polished to a deep lustre by the patina of a lifetime of use.

When a craftsman died his tools would be acquired by his remaining colleagues. Payment was made for the tools and given to the deceased man's widow, the new owner's name would be engraved, the previous name being deleted.

The screwdriver was used in the 15th century in Germany by armourers and gunsmiths. The associated craft of making screws was a Schraubendreher (screw turner) and eventually this name became the name of the tool to manipulate screws.

Every trade and every craftsman had their own ideas as to the design, shape and size of chisels and in the early 19th century suppliers would list up to 1000. One wonders whether the cost of a special chisel would be passed on to a client who demanded that extra detail in the work which only the subtle design of chisel could impart. Perhaps future work might not need the application of such a tool?

This account on hand tools has tended to concentrate on woodworking, where the evidence of craftsmanship is more noticeable. The jewellers, gunsmiths and clockmakers who worked in metal achieved high standards in decoration and manipulation of their chosen materials, but it was the woodworking techniques which were known and adapted when metalworking superseded the work of the blacksmith and then gave rise to precision engineering through the development of machine tools.

2:2 The Lathe

Lathes have been about for a very long time, and although only able to make round objects, their value is significant as turned work is more accurate than that made using hand tools.

Examples of turned work have been found from 2000 years ago and depictions of a lathe in Egypt show the workpiece in a vertical position with an assistant providing the motive power, similar to the bow drill previously described.

From this early example the lathe was developed and improved over the centuries, mainly in respect of the power source to rotate the work. Initially a tread mill was used and this was driven by a dog. The pole lathe was introduced, the power being supplied by a 'springy' hazel branch, anchored at one end with a rope attached to the other. The rope was wound around the work piece and continued to a treadle operated by the turner. As the hazel branch bounced up and down, driven by the action of the treadle, so the work rotated alternately in either direction and the turner could only cut in one direction, consequently the tool was constantly being pushed in and out. In the absence of a suitable hazel branch it has been known for a turner to use his trouser braces instead!

Another early design which did not need a treadle was a pulley and belt system, where an assistant standing behind and to one side of the turner rotated a large diameter pulley which drove a smaller pulley on the lathe. A high speed was generated using the

Development of a lathe.

ratio of a large pulley driving a smaller one and the exertion by the assistant was no doubt enormous. The belt driving the lathe was made either of catgut or hemp rope which, unfortunately, rats would eat during periods of inactivity. The external driving wheel was then incorporated within the lathe and with the treadle produced a more compact arrangement. This was the state of the art in the late 18th century. Eventually the lathe was coupled to a steam engine, driven directly or placed at the end of a line of machines with overhead shafting linking each lathe via a belt drive.

The wood turning tool, of which there are a number of shapes, e.g. gouge, skew, parting, has a long handle. When turning wood, one hand is placed at the end of the handle and is often rested on the hip, the other hand being close to the cutting edge guides the tool along the tool rest. Later, metal was cut on the lathe and with a hand held tool it became more difficult owing to the greater hardness of metal over wood. The handle in this case was extended and reshaped to allow it to lie under the arm to provide a better reaction to the greater force involved in cutting metal; it acted like a see-saw. Both hands were then free to guide and control the tool.

The shapes generated on the lathe in wood are graceful and elegant, with gentle convex or concave curves and beads and coves. With a hand held tool such profiles are slightly more difficult to make in metal than in wood, but with the advent of the metal cutting slide lathe, where the tool is fixed and only capable of moving strictly from side to side or front to back, the production of curved shapes created a problem.

In order to produce a profile on a component such as the curvature on a handle, two methods are possible – by forming or generating. In forming, a cutting tool shaped exactly to the profile to be produced is presented to the blank workpiece. A considerable force is required to cut due to the large area of contact; which could be as much as 4 inches long. Vibration and strain will be caused both to machine and the turner, the cutting tool will 'chatter' creating rippled marks on the work. It was possible to relieve the bad effects by altering the cutting speed, reducing the pressure or changing the tool geometry, but the whole process of forming was fraught with difficulty. An old attempt to cure 'chatter' was to support the tool on a pad of leather which cushioned and absorbed the vibration. Incidentally, an interesting cure for 'chatter' was applied to the boring bar (normally a long overhanging tool used to cut deep holes and notorious for suffering from vibration) by the Production Engineering Research Association (PERA) in the mid 20th century. This consisted of introducing a piece of metal (slug) into a cavity in the boring bar and allowing it to move freely. Any vibration initiated by the tool was immediately overcome and dampened by the slug vibrating in opposition to the tool's resonance.

Generating a profile is done by using a single point tool, as the turners would have done with their hand held tool. A high degree of skill is necessary to achieve the desired curvature and, to make a number of identical objects demands a good 'eye' with accurate measurements or a template to ensure success. Turning in metal by this method continued until 1800 when the copy lathe appeared which consisted of two parallel spindles – one for the blank work piece and one for the master, previously made to the desired profile. The master caused a pivoted bar to guide the cutting tool. James Watt improved the copy lathe by making it possible to make components of a reduced scale from the master, and Thomas Blanchard designed a lathe for rapid production of identical gun stocks.

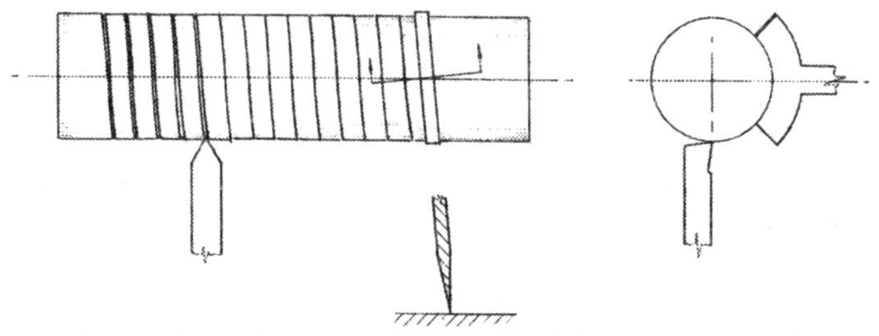

Screwcutting, where a crescent-shaped blade is set to the required angle of the thread and the cutting tool follows.

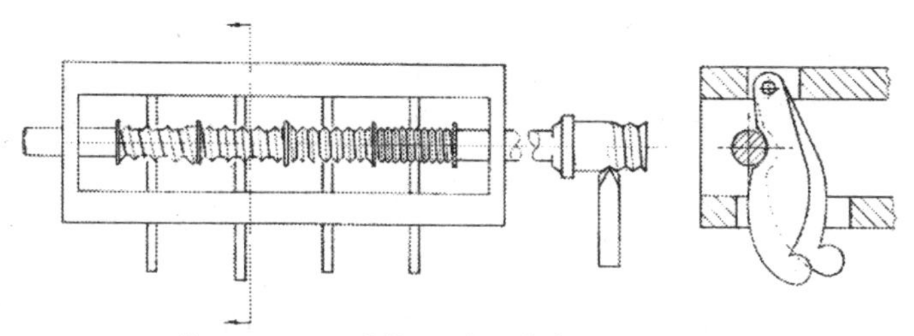

Keys engaging different threads for screwcutting.

One invaluable operation on the metal cutting lathe (now known as the centre lathe) was the making of screw threads. In the early attempts it was sufficient to relate the rotation of the work with the rate of traverse and to scratch a helical groove in the workpiece.

In the 16th century the screw mandrel was devised with a series of threads, and it was simply a matter of engaging the one required by using a key and the cutting tool would faithfully cut the thread on the workpiece.

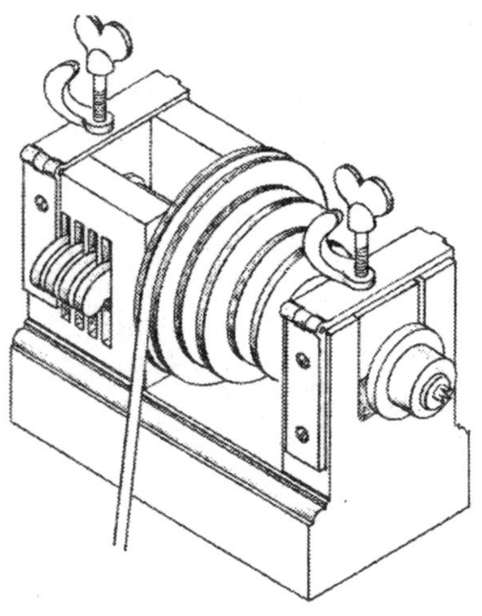

Lathe headstock with keys for screwcutting.

A coinage press was made by Cellini for Pope Clement VII [1523–1534] and the screw, which needed to be pretty substantial to transmit the force required, was constructed 'three fingers thick' with a thread of square section form. A bronze nut was made to fit by casting it directly onto the screw, having first coated the screw with a mixture of ashes and fat to act as a release agent. The contraction of the nut upon cooling allowed just the right amount of clearance for the nut to run smoothly. Whether this good fit was maintained over the whole length of the screw is not known.

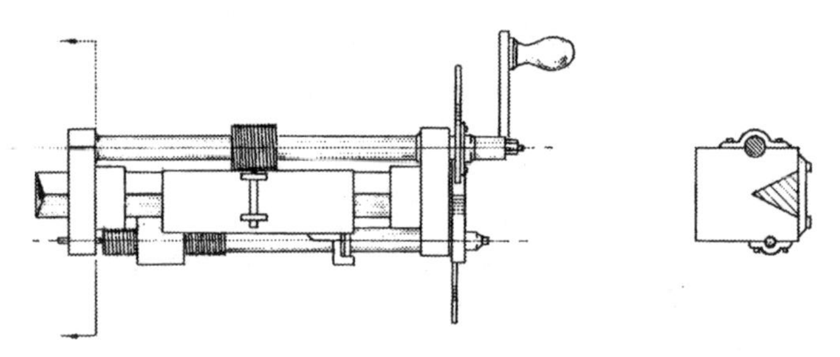

Ramden's lathe of 1773.

Jesse Ramsden [1735-1800] was presented with the problem of producing accurate graduated scales for astronomical instruments. He realised that the key to solving this problem was in the screws which controlled the marking of the graduations and so he designed and made a lathe capable of producing screw threads accurate to within 0.004in and spent 11 years perfecting this.

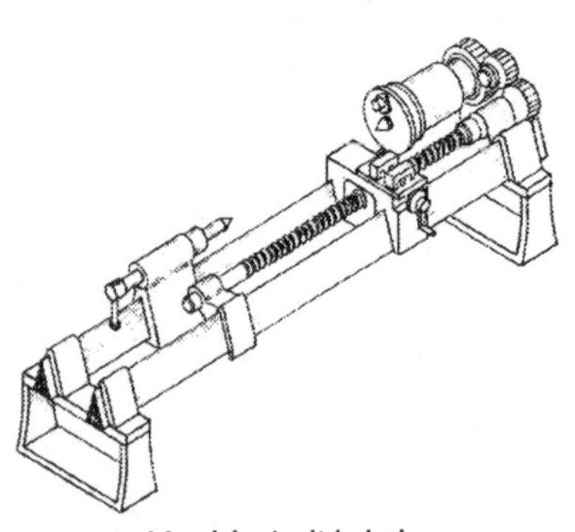

Maudslay's slide lathe.

By using such a screw to engrave scales on sextants it was then possible to measure latitude more accurately and so help navigation on those voyages of exploration during the late 18th century.

Henry Maudslay [1771-1831] may rightly be regarded as the father of machine tools and production engineering, and whilst he was working for

Joseph Bramah he improved the lathe tool slide which precisely guided the cutting tool. This was patented in 1794 by Bramah and shortly afterwards Maudslay asked for an increase in his wages of 30 shillings a week. This was refused and so he left to set up his own business which flourished and had an enormous impact on engineering. It is interesting to note that, in France, Vaucanson and Senot both came up with similar lathe designs, but Maudslay had the advantage over the two Frenchmen in that he advanced his screw cutting techniques far beyond the others.

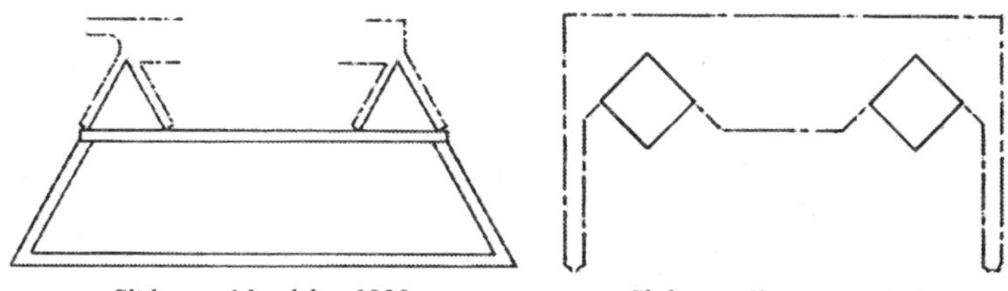

Slideway, Maudslay 1800. *Slideway, Vaucanson 1770.*

Maudslay's lathe consisted of two parallel bars of triangular section, which guided the saddle holding the cutting tool along the machine and was driven by the lead screw. At first glance this support for the saddle appears to be satisfactory but it does not take into account any change in size of the metal due to temperature fluctuations, which in Maudslay's design would cause distortion. In France, at the same time, Vaucanson designed a lathe with a similar slideway. The ideal profile is a vee and flat surface which obeys the principle known as the Three Degrees of Freedom, where the accuracy of the slideway is maintained. The base of Maudslay's lathe supporting the slide bars consisted of two castings, one at either end, which are substantial yet shaped to give an aesthetic appeal and designed to stand on a work bench.

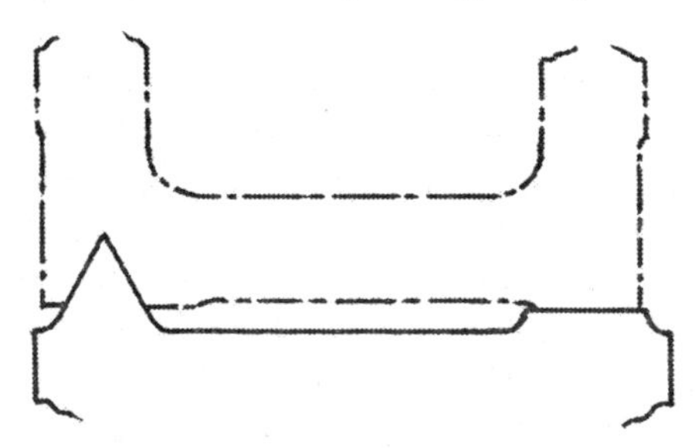

Slideway, optimum design.

The accuracy of a screw thread is only as good as the lead screw and nut of the lathe on which it is being cut, and since the lathe is the only way to make screws it would seem that the possibility for improvement was insurmountable since the lathe would consistently repeat any inaccuracies. One way of solving the problem was to use a length of cork wrapped around the lead screw of the lathe as a nut. The cork 'nut' was longer than the normal nut and any irregularities in the lead screw were equalised, therefore the screw being made was more accurate than the machine's screw. By substituting the screw made for the machine's lead screw a number of times, eventually a very accurate product would be produced.

Previous lathes were built with solid masses of wood supporting the machine, but in the early 19th century the machine tool maker had access to cast iron and exploited its potential as a substantial foundation to absorb vibration and to add style to enhance the shape. Such efforts at decoration no doubt created an appeal to a potential customer and made a statement about how good this product was, demonstrating that it was a piece of craftsmanship. The ornamentation was made by the woodworker who produced the pattern to create the mould in the sand into which the molten metal was poured. The influence for this design (or art) came from pre-existing examples, classical design and also mouldings used by cabinet makers.

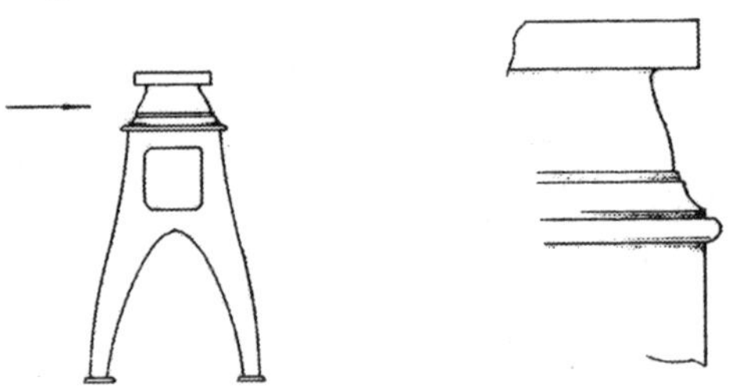

Classical influence on lathe design.

Joseph Whitworth worked on the bench for Maudslay and then went on to run his own business becoming, 20 years later, the international leader in engineering, transforming the existing technology and making immense steps forward. His particular interest was screw cutting and he realised that this was the key to precision machinery. In 1820 he made a lead screw 60 inches long and 2 inches in diameter with 50 threads per inch. Just how fine this screw thread was can be imagined by laying fine dressmaking pins side by side, a total of 3,000 to represent the length of the screw! This was clearly a landmark achievement and Greenwich Laboratory used the screw to calibrate their instruments. In recognition HM Treasury made an award of £1,000

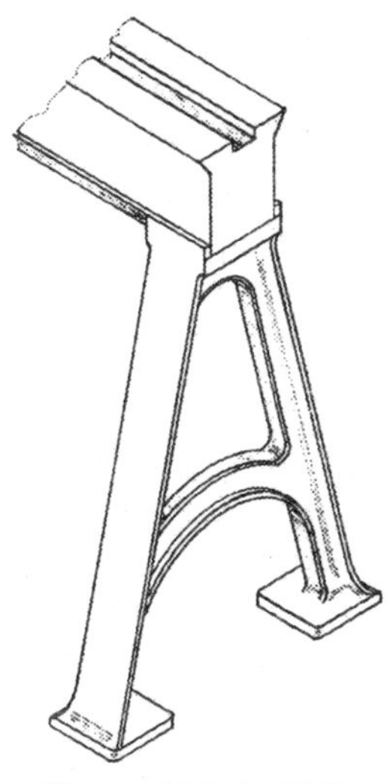
Whitworth's lathe 1843.

Joseph Whitworth's lathe comprised a bed of hollow box construction in cast iron made in 1843 and this design was still in use more that one hundred years later, with only minor changes to the method of operation.

The importance of designing lathes to the principles of ergonomics (fitting the machine to suit the individual using it) was recognised by Joseph Whitworth. He ensured that the turner, whilst standing at the lathe, would be able to operate it easily, with everything he needed within easy reach, and so work efficiently without fatigue. Sadly, there were lathes designed during the 20th century which gave little thought to ergonomic considerations and the turner would have to assume simian proportions to be comfortable whilst working.

Whitworth's first patent was for a machine to make studs and hexagonal bolts. One outstanding and unique feature of the machine was the self-centring chuck used to hold the work securely. This consisted of three jaws which moved on to the work to maintain it truly central to the axis of the lathe. These jaws were guided in a scroll (spiral) groove on the face of the chuck.

The scroll groove would have been chiselled out and filed smooth by hand, a highly skilled job demanding great precision. Quite apart from this the shape of the spiral groove would have to be decided and the calculations for this were probably worked out by William Muir, who was Whitworth's chief draughtsman and works' manager at the time. Muir himself later went on to work successfully on his own account.

The starting and stopping of lathes was achieved by moving the driving belt from a fixed pulley on an overhead line shaft to a freely revolving one. This was a simple method but highly hazardous bearing in mind that the line shaft was mounted near the ceiling and the belt would flap about around the turner's head. Another method of driving was using a clutch but early designs were ineffective. Mechanical clutch design was only achieved with the advent of friction fibre materials in the 20th century. The evidence was there in the early 19th century, when the braking of locomotive wheels was achieved by using soft iron brake blocks rubbing directly onto the periphery of the wheel, but none of the machine tool makers made the mental leap to adapt this technique to clutch design on their machines. This situation is typical of the mode of working empirically within the confines of a specialised area, and although great strides forward were made through innovation there was always a tendency to be comfortable and confident in familiar surroundings. Would a ship builder think that any advantage could be gained by talking to a coach builder?

James Nasmyth started his engineering training as personal assistant to Henry Maudslay and then in 1830 began working on his own. One of the first jobs in setting up his plant was to convert his father's treadle lathe, bringing it up to date to cut metal, and the machine was still in use 63 years later. He took out a number of patents pertaining to improvements to the lathe, including finding the centre of round bars; reversing the action of slide lathes; self-adjusting bearings and segmental work.

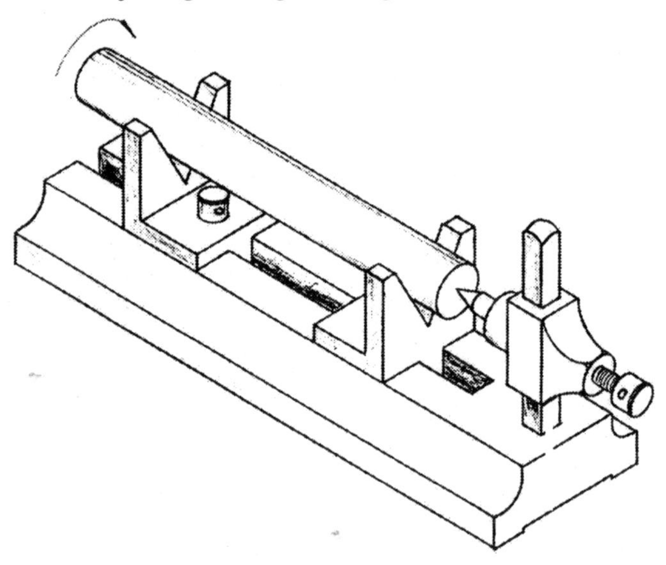

Nasmyth's device for finding the centre of a bar.

Achieving a good finish on a turned metal component required close attention to the cutting tool. It was obviously important to be sharp and to have a radius on the 'nose' so that each subsequent rotation of the work would overlap the previous cut; otherwise, if the tool were sharp pointed, a groove would be created – like a fine screw thread. Subsequent polishing using emery cloth then imparted a nice shine to the metal. If such polishing was done on poor quality turning it was known as 'deep scratch and high polish'. There are occasions where the tool mark is desirable and this then appears on the face of an item so that when light shines on it a segmental pattern is seen.

Polishing, using emery cloth (on metal) or glass paper (on wood), creates scratches on the surface caused by the grit of the polishing medium, and the way to get a smooth finish is to polish initially using the coarse material and then to progress in stages to finish with the fine grit. An exception to this is planed surfaces on wood where the smoothing plane has cut a perfectly flat smooth surface and any subsequent use of glass paper only serves to make scratches, even using fine garnet paper.

At attractive finish for steel is blueing. This involves initially getting a highly polished surface and then gradually heating the component until an iridescent (peacock blue) colour appears, it is then quenched in oil. Great care is required since if there is insufficient heat the colour is yellow, and if too much it appears black. The appearance, when properly executed, is very attractive although it is not resistant to rust.

As production engineering gathered momentum through the 19th century and into the 20th, it is worth bearing in mind that the predominant machine tool was the metal cutting centre lathe. Myriad businesses, many with only a few employees, were expert turners. The demand was so great that a skilled turner would spend his whole working life operating a centre lathe.

Machine tools were developed to replace some of the hand crafted operations. This did not reduce the skill but rather enhanced the ability and lessened the time taken in production. Machine tools can make the machines that make products and consequently they have to be extremely accurate, this therefore demands a high degree of skill by machine tool makers.

There is a story about a visitor to the site of a distinguished craftsman in wood who was invited into the workshop. On entering, the visitor was surprised to see an array of machines including a circular saw, bandsaw, sander, spindle moulder, etc.

and disappointed to find a 'craftsman' with such equipment – not realising that machines are just as much a tool as those which are hand-held.

2:3 Boring Machines

The lathes of the 18th century were only capable of turning small diameter components, but there was a great need for accurate cylinders and pistons for steam engines which were normally well over the standard lathe capacity. It was essential for a piston to fit snugly in a cylinder bore to prevent steam escaping and ensure an efficient engine.

Creating a perfectly round cylinder bore which could be up to 63 inches in diameter was impossible in the early 18th century and the engine builders endured the best that could be provided using hand tools. The Dutch had bored cannon barrels from a solid blank of metal but a steam engine cylinder bore being much longer would be rough cast as a hollow blank. John Wilkinson patented and built a boring machine for cannon barrels and engine cylinders in 1774. It consisted of a rotating bar along which the cutting tool was moved and the workpiece was

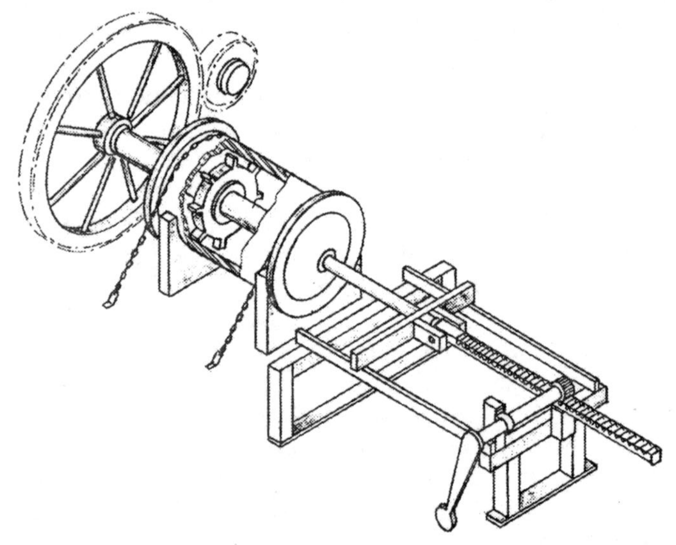

Wilkinson's boring machine.

clamped firmly to the machine. The degree of accuracy achieved on a 50 inch diameter cylinder bore was such that the gap on assembly with the piston was no greater at any one place than the thickness of a one shilling coin (0.020in). John Wilkinson was primarily an ironmaster producing iron castings. His determination and ruthlessness gave his company a wide reputation for the complete production of cylinders, cast and finish machined.

Other engineers contributed their expertise to improve boring machines, including Matthew Murray and Bodmer, and better quality finishes were achieved, with greater accuracy on the machining of the flange faces square to the bore.

Early methods attempting to create circular bores in cylinders, before the advent of boring machines, were crude yet innovative. Richard Reynolds, in the early 18th century, described his method of 'scouring' the bore of a cylinder 28 inches in diameter and nine feet long where he allegedly cast lead in a semicircular shape, this lump was then smeared with emery and oil and pulled to and fro in the rough bore by a number of men. By rotating the cylinder little by little the whole of the bore was made smooth and the degree of roundness was such that the difference between the greatest and least errors in diameter was less that the thickness of Reynolds' little finger. This was quite an achievement for the method employed; one wonders how long the operation took to generate such an accurate bore using the measuring instrument of the human digit.

2:4 Planing Machines

In engineering the need for flat pieces of metal was not as great as the requirement for round components. The long history of the lathe reflected this with the need for items for mechanisms which invariably turn round and round, with a flat surface only needed to support these parts, and that was normally a flat (rolled) sheet of brass, for instance the end plates supporting the mechanism of a clock. As machine tools began to be developed it was essential that a perfectly flat bed was made to guide the tool along a straight path.

Hence planing machines were invented and, co-incidentally, such machines could make themselves. The principle of operation was that a cutting tool makes successive cuts along the work piece, each cut being taken alongside the previous one until the whole area was covered. The work piece could be up to 12 feet long. The earliest planing machine was designed and made by Richard Roberts in 1817 and incorporated a device to lift the tool on the return stroke thereby preventing rubbing and extending its life. Two distinct types of machine were made, one where the work is stationary and the tool is attached to a gantry running on flanged wheels between two parallel rails. The other type is where the workpiece – usually a massive block of cast iron – moves along the machine under the stationary cutting tool.

Planing machines were essentially used to make long components like machine slideways or bases but where a flat surface was required on smaller items the shaping machine was used. This consisted of a vice to hold the workpiece over which the tool reciprocated back and forth, moving slightly sideways at each cut in the same manner as the planing machine. The tool was driven by a crank attached to a rotating shaft. It took a long time to create a flat surface on even a small workpiece so Joseph Whitworth devised an ingenious mechanism to speed up the tool on its return stroke thereby increasing the efficiency of the

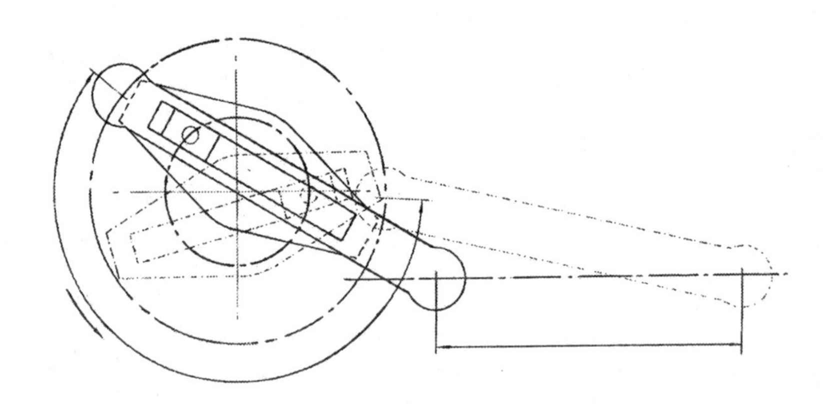

Whitworth's quick return mechanism. A constant speed of rotation adapted to give slow cutting stroke and a rapid return of the cutting tool.

machine. This was known as the quick return mechanism. A greater innovation was the reversible tool, where it was arranged to cut on the return stroke. This device was known as a 'Jim Crow'.

In time it was recognised that the planing machine had certain limitations as far as the creation of more complex flat surfaces on components was concerned, such as slots, grooves, recesses etc. The answer to this was the milling machine and one early example is the machine designed by James Nasmyth in 1829 to cut flats on a round bar to make hexagonal nuts and bolts. It consisted of a table holding the workpiece which was indexed around to present it to the rotating cutting tool. An entirely new concept was born and the only link to planing and shaping machines was the production of a flat surface. Nasmyth went on to make a machine to specifically cut the slots for drive keys in wheels and pulleys which were fixed to line shafting.

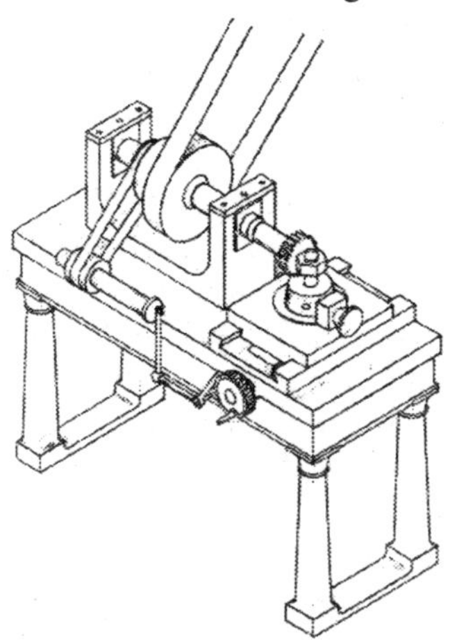

Nasmyth's nut cutting machine, 1929. An iron table with legs reminiscent of classical Tuscan pillars.

The base of a machine tool needs to be substantial, stable and absorb the vibration resulting from the cutting action of the tool. Nasmyth's hexagon cutting machine has a table carrying the work and cutter supported by four elegant legs reminiscent of classical columns with capitals and bases. Machine tool design eventually became more robust but Nasmyth's wish to create something which to our eyes is attractive can only be admired.

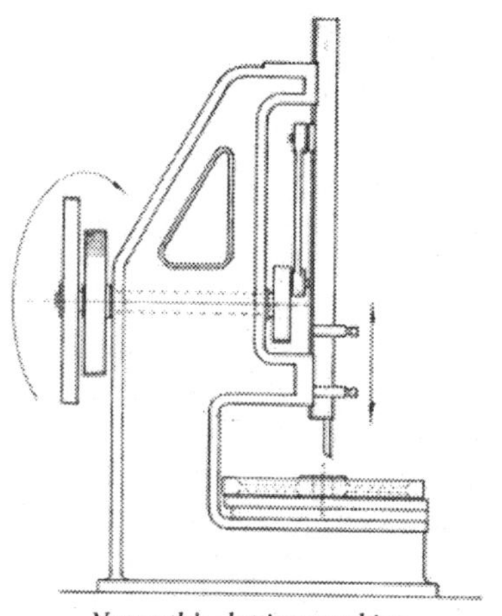

Nasmyth's slotting machine.

2:5 Forming and Forging

The processes described previously are all concerned with the removal of material, 'wasting' to make the desired object. The method used to reshape the raw material to produce the component in metal is generally known as forming, forging or stamping.

To change the shape and deform metal takes a great deal of pressure. One very early example is the screw press and although primarily used for paper, olives, grapes etc, rather than metal, the principle is the same. The vertical columns joined by a cross beam supported the screw which delivered the pressure. Such a press was capable of great force and worked by turning a handspike. A screw with a thread of one inch pitch and a handspike three feet long with the operator applying a 40lbs force gives a pressure of 9,000lbs at the end of the screw.

Coins have always been made by the process known as stamping, where the negative form of the image required is imprinted into the tool called a die and this is used to make the coin from the blank workpiece by delivering a sharp blow. The making of the die demands great care and must be a faithful copy in detail from a plaster of paris model. Apparently in 1830 Henry Maudslay perfected the technique and this appears to be the forerunner of the present-day die-sinking machine.

No forge in England or Scotland was capable of making the paddle shaft for Brunel's SS *Great Britain* which was 30 inches in diameter. This was due to the throat, the distance between the hammer and the workpiece, being too small and, therefore, insufficient momentum was achieved to make any impression on the work. Such a situation was described as the forge being 'gagged'. James Nasmyth came to the rescue with his steam hammer in 1839; however, the SS *Great Britain* was fitted with a screw propeller and the large paddle shaft was not required. At that time there was a recession in the industry and no one wanted the new steam hammer. Then some visitors from Le Creuzot, in France, were shown the drawings of Nasmyth's steam hammer and on returning to France they built one. This had imperfections which were later corrected by Nasmyth. He was reluctant to take out patents for his inventions, believing that the wide adoption of his ideas would be beneficial and of service to industry as a whole, quite apart from the cost and trouble of toiling through the bureaucratic process of patent application. However, on this occasion, due to the French advantage he did patent it. The steam hammer had a very sensitive control and it was possible to just crack an egg or, conversely, deliver a blow causing earth vibrations two miles away. A steam hammer was patented by James Watt in 1784 which worked by using exhausting steam from a steam engine releasing the hammer rather than being direct acting as in Nasmyth's design.

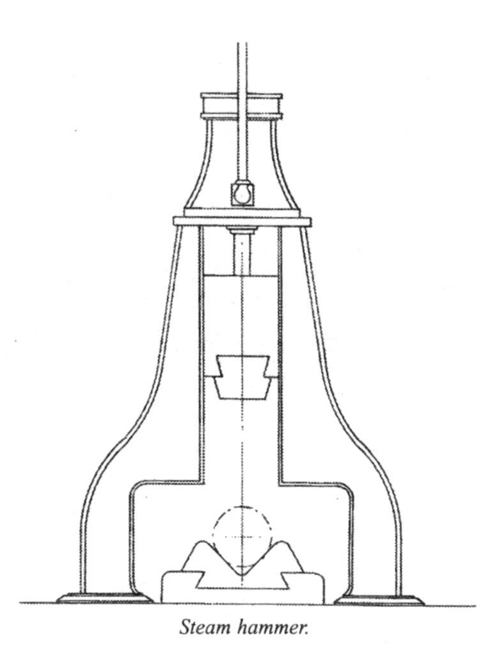
Steam hammer.

Ships' anchors were made piecemeal, welding and hammering by hand, and often broke in service due to the high demands placed upon them. Welding by hammering had the advantage because the impurity of slag is driven out, but forming from one piece using the steam hammer transformed the crystalline structure of the iron into a linear form, like the grain in wood, and made the anchor stronger than the alternative welded construction.

The helve, and the lighter tilt hammer, are medieval machines used to forge and shape iron as an alternative to hand hammering by the blacksmith. The helve is driven by a rotating shaft having a number of cams protruding from the circumference which in turn lift the tail of the beam holding the hammer and then allow it to drop quickly, thereby delivering a sharp blow to the work by the hammer. The helve was a crude yet effective tool in forging, the hammer weighing up to eight tons and acting under the force

Tilt hammer.

of gravity. With four cams lifting the hammer on a shaft of seven feet diameter, rotating at 15 revolutions per minute, 60 blows were delivered per minute. The working faces of both the hammer and the anvil upon which the workpiece rested were plates of wrought iron and easily replaced when worn. During the operation great skill was required to gauge the exact position of the work whilst the hammer was falling every second. Care was also necessary due to the craftsman's fingers being very close to the hammer.

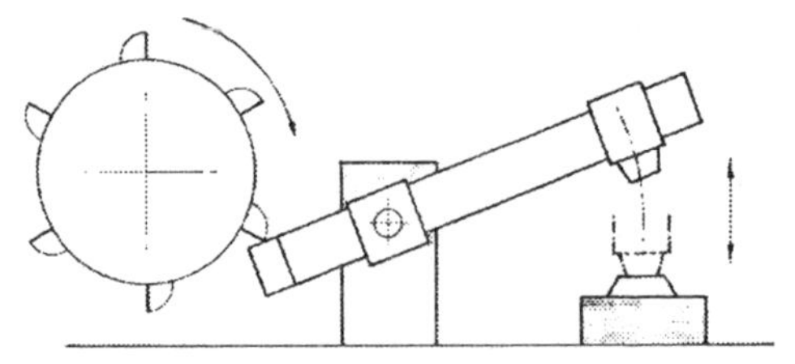
Principle of the helve or tilt hammer.

In addition to the methods of forming or forging already mentioned there is hydraulic power. This is mainly used to produce pressure rather than to strike a sharp blow. Joseph Bramah made a hydraulic press in 1795 to extrude lead or tin to make pipes. One problem with any hydraulic system is the seal between the piston and the cylinder through which it passes. If the seal is too tight the piston does not move and if too slack the effectiveness of the system is lost. Joseph Bramah had used hemp packing as a seal and at that time Henry Maudslay was working for him. Maudslay noticed the inefficiency and came up with the cup washer made of leather. This device was self sealed against pressure and offered little frictional resistance on release. This survived into the 20th century, the bicycle pump being a familiar example where the flexible cup washer is able to adjust to any imperfections in the bore through which it slides, providing pressure on the push stroke and drawing in air on the return.

The processes described in connection with the forming, forging, stamping and pressing of iron were all improvements on the previous hand crafts performed by the blacksmith. With mechanisation it was easier to make very large work pieces. However, the blacksmith was still needed for the smaller items as a jobbing producer and to make decorative items using small bars of steel both round and flat which he would bend, twist, pierce and hammer-weld into exquisite artwork. The blacksmith, by definition, worked at a forge manipulating hot iron. In contrast the whitesmith worked with cold metal, generally sheet materials of various metals, to make decorative artwork and on a commercial basis. Both blacksmith and whitesmith practise today predominately in the art and craft domain.

3
Products

3.1 Gears

Gears are wheels which have teeth and these can project externally, internally or from the face of the gear.

Gears are sometimes referred to as cogs but these are normally associated with the bicycle where the cogs engage the chain. Gears are used to transmit motion and to change the rate or direction of rotation between parallel shafts or those inclined at an angle. Thus, a gear mounted on one shaft is meshed (engaged) with a mating gear.

To ensure smooth operation and the efficient transmission of power the shape of the sides of the teeth is important. Ideally it should be a curve which a point on the arc of rotation would travel when in contact with a point on a mating tooth. However, there are exceptions and examples are found where the teeth are of a more simple shape.

Typical gear train.

Without delving into the geometry of tooth design it is sufficient to say that there are two forms that are generally used; the involute and the cycloidal; both of which satisfy the requirements but are based on different principles. The involute curve is described by the end of a taut line as it is unwound from the circumference of a circle. The cycloidal curve is generated by a point on a circle as it rolls along a straight line. The cycloidal form of tooth is used by the clockmaker. The superiority of cycloidal teeth in the gearing for clocks was first shown by Roemer and Huygens in 1674. Smeaton subsequently found by experiments on models, that uniform motion was best achieved using the cycloidal profile on gear teeth. It is interesting to note at this point that in the case of clocks the larger wheel drives the smaller pinion producing an increase in speed, i.e. geared up. On the other hand, in general engineering, the reverse applies, where a motor or prime mover reduces the speed and increases the power, so the involute tooth profile is used on gears where high power transmission is required, e.g. on machine tools and heavy duty machines.

Clock making benefited from the scientific study of tooth form but the heavy duty gears required to work the watermills and windmills were ignored until well into the 18th century. The clockmakers had a long tradition of needing to achieve consistent quality and they became somewhat elitist. This situation was the outcome of demands from their clients who sought to acquire the very best of the current fashion, whether it was for the engineering or the aesthetic appeal. The wealthy would commission the humble craftsman to produce a timepiece and he would strive for excellence; as a result, in time, his humble status would be exalted.

Gear teeth of triangular form.

Before the benefits of the tooth form was realised in the 18th century, the very early clocks had gears with triangular shaped teeth. Much skill was needed to ensure that all the teeth on such a gear were identical and equi-spaced, and also on the mating gear. Using only hand tools great accuracy was achieved since the slightest misalignment would cause the clock to stop.

The precision engineering of a clock mechanism is often admired, since it contains many gears and, whilst accuracy is vital, it must be realised that freedom of movement throughout the gear train is equally important. Clearance is provided between the gear teeth and also where the spindles (shafts) carrying the gears enter the end plates. The very slightest pressure with a finger on the minute hand is sufficient to stop a clock. The power to drive a clock by either weights or a spring is initially very high and only decreases through the gear train; the size of the gear teeth reflect this and diminish accordingly.

The most commonly used gear is the one having teeth projecting from the periphery and known as the spur gear. Producing teeth to the correct profile presented problems in perfecting before the advent of specialised gear cutting machines. Both the involute and cycloidal forms can be approximated to arcs of circles and this made life easier for the woodworker marking out teeth for wooden gears or for a pattern in wood used to make cast iron gears. There is a great variety of combinations of tooth size, number of teeth and diameters, so the craftsman would have to be aware of the requirements for power transmission and size for any particular job. A device to make life easier in calculating a tooth profile was the odontograph, which gave the position of the centres of radii to approximate the form based on tables of data and scales.

James Brindley, in 1775, was involved in the building of a new silk mill in Cheshire and had to make gears for the operating mechanism. He did not want the teeth on the wheels cut by hand, which was a time-consuming operation with the possibility of poor quality, so he designed and made a gear cutting machine which had a rotary cutter shaped on each side with the negative profile of the tooth, with the result that the silk mill operated efficiently and endured. Brindley found that his workmen always stopped work promptly at 12 noon but were not so punctual to return at 1 o'clock. Their excuse was that they could not hear the clock strike; so apparently Brindley adjusted the clock to strike 13 at 1 o'clock!

Some forty years later Richard Roberts improved the gear

cutting machine and made the Sector. This was a gauge to establish the correct number of teeth for a given diameter, or vice versa, and ensured that the final cut did not leave half a tooth. His gear cutting machine was capable of cutting any number of teeth on gears up to 30 inches in diameter. The previous method of making the gear complete including the teeth by casting in iron had the disadvantage of creating a rough finish and the sides of the teeth had to be filed smooth by hand. There was also the possibility of poor quality due to distortion and cracking as the metal cooled down. However, one advantage in cast iron is that it has a hard skin and would therefore resist wear due to abrasion.

In medieval times it was sufficient to position protruding wooden pegs from the edge of a wheel and allow them to engage with a similar arrangement on the mating wheel. Since the teeth were crude wooden pegs they would eventually 'bed in' and then assume a profile closely approximating an involute curve. In so doing they would be reduced in size and less likely to be able to transmit a high power without fracturing.

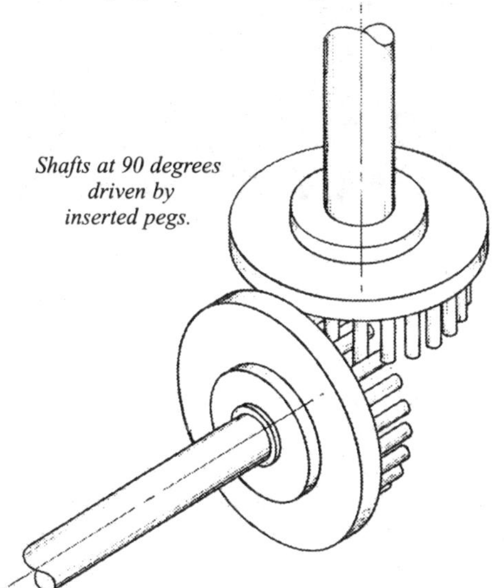

Shafts at 90 degrees driven by inserted pegs.

The noise caused during the bedding in process by wooden gear teeth would no doubt be deafening and not much reduced afterwards, when excessive clearance and backlash between the teeth would create rattle and vibration. Typical examples of such gears can be seen in windmills and watermills and there is often evidence of teeth having been replaced due to fracture. A close grained wood is necessary to ensure a reasonable life of a tooth, and apple, holly and hawthorn were favourites.

Gears made entirely of wood in watermills eventually gave way to the construction of the wheel structure in iron and wooden teeth inserted into sockets around the periphery. Such a gear would be meshed with a smaller gear to change the speed of rotation and this smaller gear, known as a pinion, could be made entirely in iron. The advantage of meshing gears with teeth of different materials was quieter running and less wear.

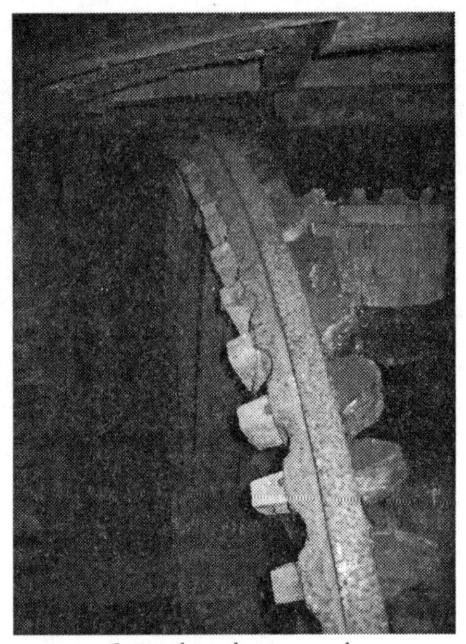
Inserted wooden gear teeth.

Eventually, when the tooth profile was understood and could be cut, the teeth of both mating gears were iron, but the gearing in watermills with metal teeth meshing with wood remained, and the noise level was then generated more by the creaking of the timber structure than the rattling of ill-fitting gears.

It was recognised that having the gear on the waterwheel shaft placed a great stress on the spokes and so the waterwheel was re-designed by putting gear teeth on the inner edge of the outside of the waterwheel and connecting this 'internal gear' to the pinion driving the mill. The great advantage gained was that the spokes could be made much lighter – in tension rather that compression. The driven pinion was also positioned close to the toothed waterwheel and overcame any torsion or twisting previously caused by having it at the end of a long shaft.

This improvement was only really possible with the ability to cast segmental gears – a strip of teeth around the arc of a circle – and a number of such segments attached to complete the full gear wheel. It was in 1769, when Smeaton was the consulting engineer to the Carron Ironworks in Falkirk, that the all metal waterwheel was made and marked the end of the wooden construction which had lasted for eighteen centuries.

One gear which departs from the convention of having teeth is

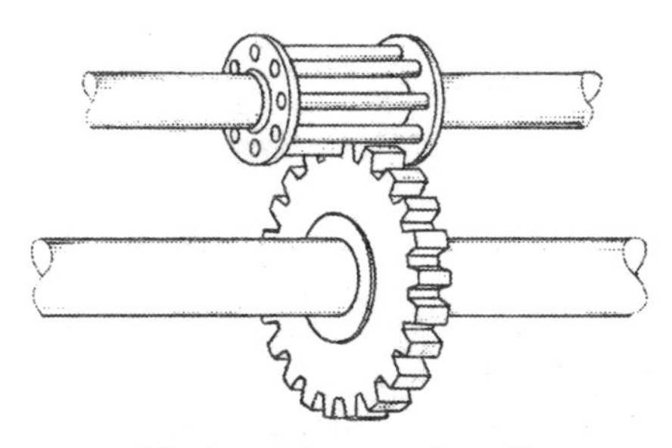

A lantern pinion engaging with a conventional gear wheel.

the lantern pinion. This was used in clocks and watermills at a point in the gear train where power requirements were low, and the teeth of the lantern pinion consisted of circular spindles (trundles) which mesh with a mating gear. Under light loads the need for a special tooth profile was unnecessary and the arrangement worked well, friction was low and when trundles became worn or broken they were easily replaced by detaching from the end plates (shrouds). In clock movements, where small lantern pinions were used, the trundles were sewing needles made of hardened steel.

Lantern pinion in a clock movement.

Using gearing to change rotary motion into linear motion is achieved using a rack and pinion. The pinion teeth are normally to the involute profile whereas the rack would have straight sided teeth with identical spacing to those on the pinion. One application dating from the early 19th century is the mechanism used to raise and lower the paddles on canal lock gates. The pinion attached to the winding handle is of a small diameter with only about six teeth which are large and placed in a staggered double row.

The rack is similarly configured and in operation there are always two points of contact on adjacent teeth. An interpretation of the reason for this arrangement is that the large teeth were necessary to carry the load, especially since they were made in relatively brittle cast iron and liable to fracture if of smaller proportions. Also, by staggering the teeth it was easier to operate the handle since the effective tooth size was less coarse.

Considering that this mechanism was made when the inland

waterways were being constructed in the early 18th century, it is a remarkable achievement in terms of design and manufacture, bearing in mind the limited techniques available at the time.

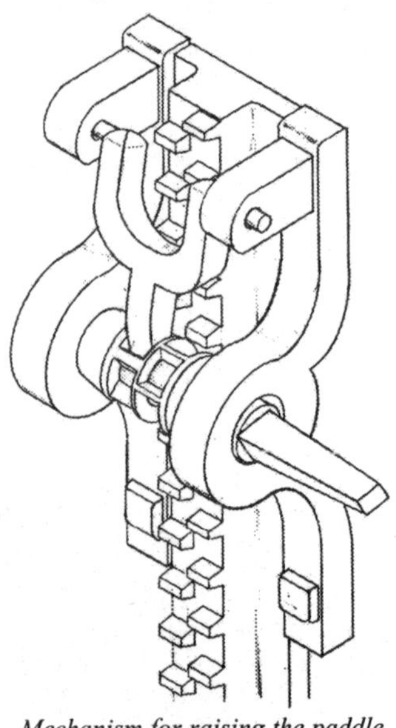

Mechanism for raising the paddle on a canal lock.

View of a canal lock.

An unusual example of the use of a rack and pinion in a house dating from Georgian times is the means of securing the bolts on the large and heavy main door of the house. A pinion engages with two racks, one on either side, and the pinion is rotated by moving one rod up or down using the hand-knob. The other rod then moves simultaneously in the opposite direction. So, as the upper rod enters the socket in the lintel above the door, the lower rod penetrates the threshold. This neat system overcomes the problems of conventional methods where long vertical bolts tend to drop under their own weight and also demand effort in operating.

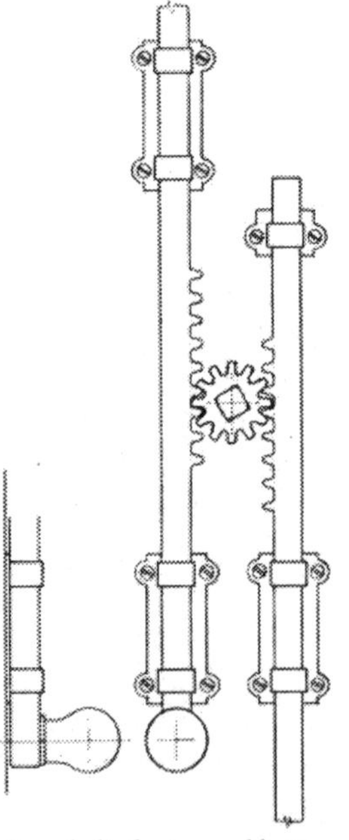

Door bolts for top and bottom operated simultaneously.

When a gear train operates in one direction without reversing, wear will occur on only one side of the gear teeth. After a long period of service under

Wear on one side of gear teeth after long service. Almost half the thickness of the teeth has been worn away.

heavy load the flanks of the teeth receiving the pressure become deformed. The question then arises as to whether it would be cheaper to make gears with the driving side of the teeth to the involute form and to leave the other side 'as cast'.

During the late 18th century there was a need to improve beam engines to make them drive a wheel instead of being confined to work only 'up and down'. This was solved using the crank and connecting rod principle and James Watt, although credited with the idea, found that it had been patented by a competitor. He then proposed alternatives and patented five other methods in 1781. one of these was the 'Sun and Planet' gear which was invented by William Murdock whilst in the employ of James Watt. In operation it works by the sun gear being attached to the shaft of a flywheel and the planet gear, which meshes with the sun gear, connected via a connecting rod to the beam of the engine. So it travels around the sun gear as the beam moves up and down

Rotative beam engine.

and causes the flywheel to rotate.

If the sun and planet gears have the same number of teeth the rotating flywheel will make two revolutions for each stroke of the beam engine. It is worth mentioning here the purpose of flywheels in engines. A flywheel is always a heavy mass of metal and possesses inertia, it demands extra power to start but once going it is difficult to stop. When rotating it tends to make an engine run more smoothly and once the fuel supply (steam) is removed the engine will then gently come to rest.

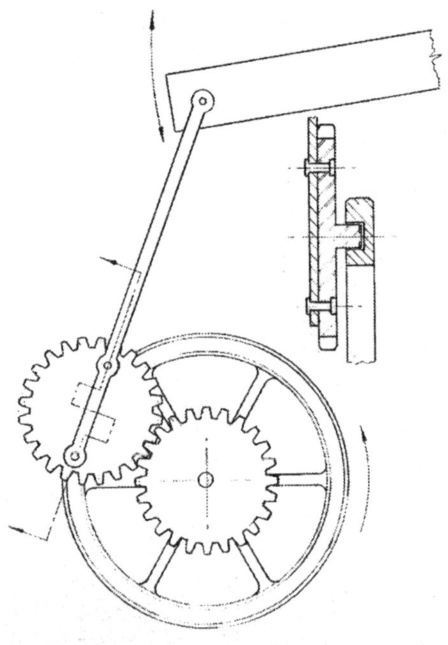

Watt's Sun and Planet gear, 1794. Oscillating action of the beam is transformed to rotary motion of the flywheel.

The sun and planet gearing was used on some of the early bicycles. Sturmey was apparently extremely pleased with the mechanism. The pressure on the pedal (reciprocating motion) via the crank drives the planet gear round the sun gear attached to the wheel (rotary motion).

One simple yet ingenious method of changing reciprocating motion to rotary motion is where a segmental pinion engages alternately with a double rack attached to a fork shaped bar. Further, if the forked rod is reciprocated (moved backwards and forwards) it causes the pinion to rotate.

With the advent of the rotative beam engine and the need for accurate large gears the trade turned to the millwright, who was familiar with large gearing in watermills and windmills, and it was natural for him to be 'upgraded' to satisfy the new requirements. Even so, they did not achieve the status of the clockmakers, although the term millwright continued into the

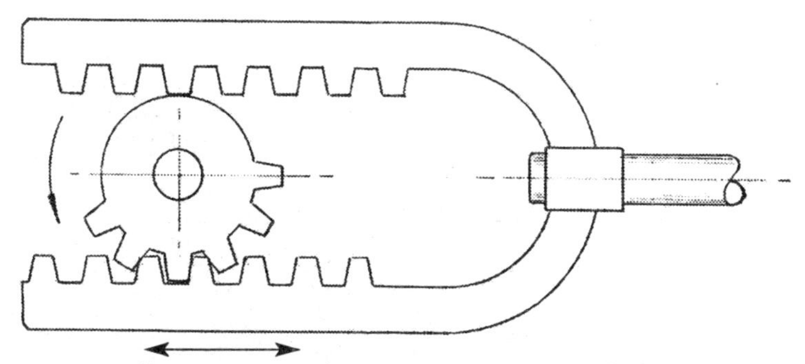

Cutaway pinion produces reciprocating motion to the fork.

20th century to be applied to those men who were responsible for the movement of heavy machinery often demanding great care with precision equipment.

Robertson Buchanan, writing in 1823, produced *Practical Essays on Mill Work and other Machinery*. A document containing a wealth of advice and information, surprisingly advanced considering the state of the art at that time. Included is his calculation for the thickness of teeth in cast iron gears: 'Find the number of horses which are equivalent to the power of the first mover of the train of machinery, and divide that number by the velocity, in feet per second, of the pitch line of the pinion or wheel; extract the square root of the quotient and three fourths of this root will be the least thickness of the tooth for the wheel or pinion in inches.' Interpreting this statement and making some assumptions we can make some sense of it, e.g. 10 horses divided by 3ft per second equals three and one third, the square root of this is 1.83 and three quarters of this is 1.45 inches. So a tooth thickness in cast iron of just under one and a half inches will transmit 10hp. He also pointed out that 'the kind of wood employed for teeth is normally about one fourth of the strength of cast iron, and since the thickness of the teeth should vary inversely as the square root of the power of the material, the square root of one quarter being one half, wooden teeth should be twice the thickness of cast iron teeth.' This was a remarkable account and contained such esoteric language it is surprising, bearing in mind the relatively high illiteracy rate amongst craftsmen at the time, that anyone would be able to make gears in accordance with the theory.

One application of a quadrant – gear teeth only on part of the periphery of the wheel – is the location system for the bascule bridge on inland waterways. The quadrant is attached to both sides of the cantilever bridge and engages with a rack on the ground. The bridge functions effectively despite the ravages of the weather and is more reliable that the alternative of using a shaft and bearings. Some of the bridges have a series of holes on the quadrant which locate in matching dowels protruding from the ground. This is not as successful (although cheaper to make) since the holes become packed with debris and prevent accurate location on the dowels, thus causing misalignment when the bridge is lowered. Incidentally, the word bascule is French for see-saw. An apt description of the operation of the bridge, where the weight of the bridge deck is counterbalanced by an extension on the landward side. Such bridges are common on the many canals in the Netherlands.

Bascule bridge over a canal.

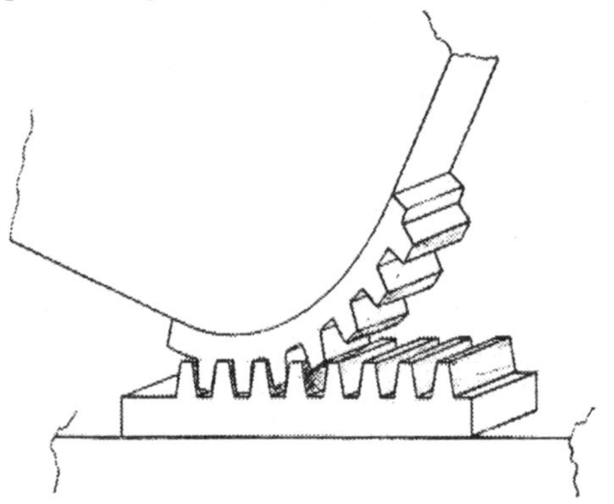

Gear quadrant attached to the bridge engages with a rack on the ground and acts as a pivot.

The worm and worm-wheel is a different type of gear from those previously mentioned and is used where a significant reduction in the speed of rotation is required between two shafts at 90 degrees. The worm needs a small amount of force to rotate the wheel. The device was used in ancient times and is of value where a load can easily be moved without danger of the worm-wheel reversing when the effort is released.

The alternative to the worm and wheel drive for shafts at 90 degrees is the bevel gear, having conventional gear teeth set on the face of the wheel at an angle. The crown wheel and pinion has a similar purpose but with a conventional pinion driving a wheel with teeth on the edge of the face. Originally the crown wheel and pinion would consist of wooden projecting pegs.

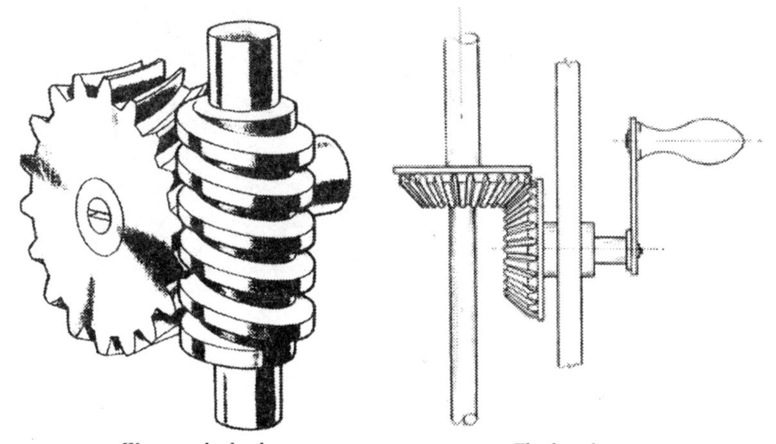

Worm and wheel. *The bevel gear.*

One of the most complex and sophisticated astronomical instruments, a geared calendar known as the Anikythera, was discovered in the Mediterranean Sea and dates from about 80BC. It consisted of more that 30 bronze gears with triangular shaped teeth attached by pins and wedges. One of the gears in the mechanism had 27 teeth on a diameter of four inches, this was not repeated until ten centuries later in clock making! The oldest evidence of Islamic and Chinese astronomical machines was 1,000AD.

Where a pinion is driven by a handcrank to drive a gear train on a rack, such as a winch or canal lock paddles, there is a tendency for the gear train to reverse when the handwheel is relaxed. This can be frustrating and dangerous and is overcome by using a ratchet and pawl. Sometimes the ratchet was dispensed with and the pawl rested, under gravity, directly on the gear teeth. This was possible with coarse gearing where secure contact could be made. The pawl swinging from a pivot point had to be loose and on outdoor locations the wear, within limitations, made it work more effectively. The alternative name for a ratchet and pawl was 'click' and this term was used by clockmakers due to the sound it made as the clock was wound, with the pawl held against the ratchet by a light leaf spring.

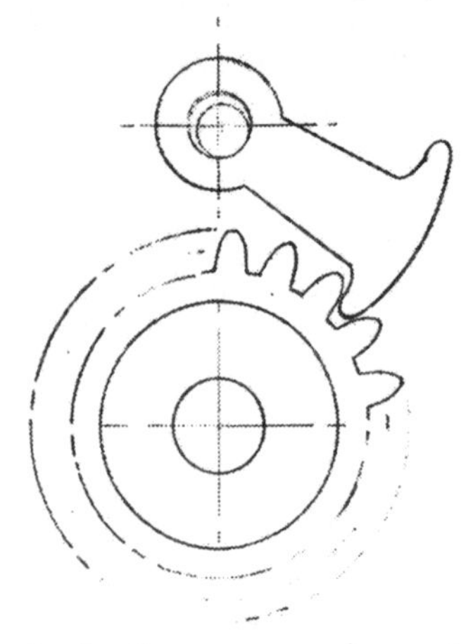

Pawl resting on a gear under gravity to prevent reversing.

Ratchet and pawl on a clock, known as a 'click'.

3:2 Wheels

With the exception of gears, wheels fall into three categories:

1. Those which, in contact with the ground, provide the means of transportation
2. The hand wheel used to control machines etc.
3. The wheel as an integral part of a mechanism; e.g. the water wheel.

The wheelwright was the craftsman entrusted to produce wheels for wagons and carts, and over a period of many years the ancient art was refined to perfection.

Each part of the wheel was made from a material having the properties best suited for the purpose. The hub (nave) was elm, a wood very dense and resistant to water. Iron bands were placed around the nave to prevent splitting. The box, made of iron, was inside the nave. The spokes were oak, a hard, dense wood suitable for load bearing. Ash was used for the rim (felloes – made in segments) due to its spring-like nature and the ability to absorb shock. Around the rim an iron hoop was tightly fitted as a tyre.

Attaching the iron tyre to the completed woodwork involved heating it to red hot and then pressing it onto the rim, the contraction on cooling would then secure it. The tyre would have been made from a strip of iron, the length calculated accurately to ensure that when made into a hoop, a tight fit on the wood was assured. There is some evidence of cart users driving their vehicles through shallow ponds in very hot weather to secure a tyre which had become loose, the water caused the wood to swell and tighten against the iron tyre. This procedure should not have been necessary if the tyre had been fitted correctly in the first place. Ideally it should have been done on a hot day when the wood was dry and the tyre a tight fit when cold, the expansion, when red hot, providing the slight clearance to allow fitting.

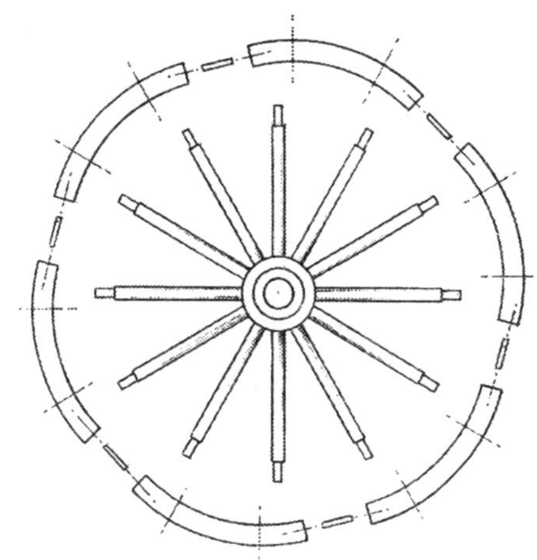

The cartwheel components.

The spokes, where they are jointed into the nave, were alternately staggered to increase the amount of wood around the joint and avoid splitting of the nave. The box was hammered into the centre of the nave and the wheel rotated on a spindle to check that it was running true, without wobble; wedges were then driven in to secure the box when the wheel was truly concentric. When finished the wheels were painted red or given a coat of tar to keep out the weather and to prevent attack by insect or fungus.

A cart or wagon wheel is made slightly dished, i.e. the nave is offset from the rim. This is to counter the sideways movement due to the motion of the

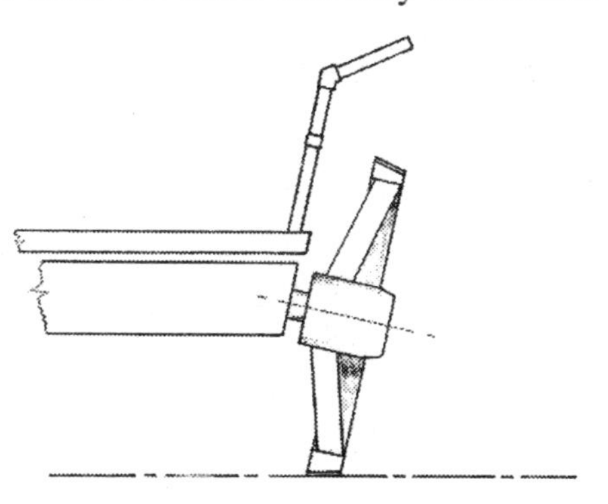

Cartwheel showing the effect of dishing.

horse drawing the wagon. Dishing of the wheel resists this motion and it is impossible for the spokes to bend due to the diaphragm effect. An advantage of the dishing is that a wider load can be accommodated on the wagon since the uppermost part of the wheel projects more than the nave. Also, the spokes below the nave are always vertical as the wheel revolves and, therefore, better able to support the load. The wheel is mounted on a stub axle which is canted downwards to complement the angle of the dishing. This has two advantages; it provides greater ground clearance by raising the body of the wagon and is also pointed forward to give a 'toe in' and so makes the wagon easier to steer or pull in a straight line and prevents 'wander'.

These observations on the design of the humble cartwheel reflect a reasoning process and development passed down by word of mouth over many generations without recourse to drawings or scientific data.

Prevention of a cartwheel sliding off the stub axle was achieved in various ways. The remains of a Bronze Age chariot showed an ingenious yet effective method involving a cowhide strip located in a slot in the axle against a washer which rubbed against the rotating wheel. Much more recently a pin or headed key was driven through the axle. Such a fixing is usually concealed under a collar which extends from the hub and at first glance it appears to be impossible to remove the key or pin when changing the wheel, due to the limited clearance. However, a small slot is provided in the collar to facilitate removal. In time, with the development of screw threads, the castellated nut secured by a cotter pin became the norm.

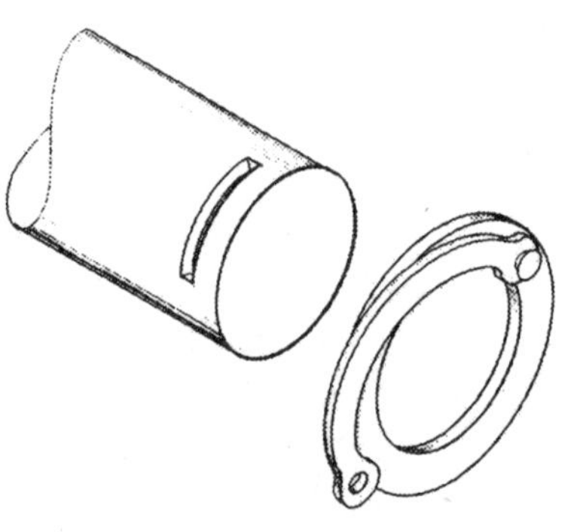

Bronze age wheel retaining disc.

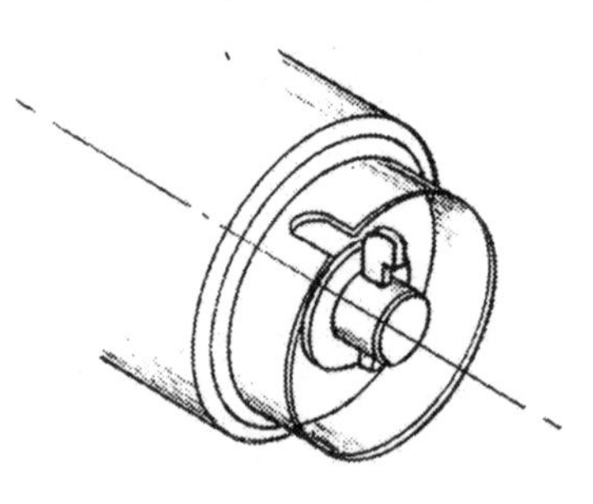

Key retaining wheel on an axle.

The all metal wheel used to power or control machines was normally made of cast iron and those used on agricultural equipment were particularly attractive with their graceful sweeping spokes. By using cast iron, such wheels were simple to make and cheap to reproduce in quantity once a wooden pattern had been made. The pattern could contain much intricate detail giving the woodworker an opportunity to demonstrate his talent whilst still satisfying the functional requirement. The pattern-maker had responsibility to ensure that metal was cast in the right places. Where strength was needed the metal was thicker, gussets and lugs were provided and overall he tried to create uniformity so that the casting would not crack or distort on cooling. All the features in a wooden pattern were faithfully reproduced in a metal casting and where a hand wheel is attractive and invites contact it would be a pleasure to use. If it looks good it will feel good. Combining ergonomic requirements and function with aesthetics is excellent design.

One evident difference between the cast iron wheel and the cart wheel previously described is the curved spokes. It could be argued that when an effort is applied to the rim of the wheel to turn the hub the resistant force would tend to bend radial spokes and by having curved spokes this problem would be overcome. However, if there was a need to reverse the direction of rotation such an advantage would be lost.

Curved spokes on a cast iron wheel.

The real reason for the curved spoke is due to the casting process, where the metal, on cooling, contracts and with a radial spoke configuration there would be cracking and distortion of the rim of the wheel but with the curved spokes the metal simply twists slightly. On the agricultural machine used for cutting chaff the blades are fixed to the curved spokes of the wheel. By doing this a shearing cut is made which demands less effort than presenting a flat blade.

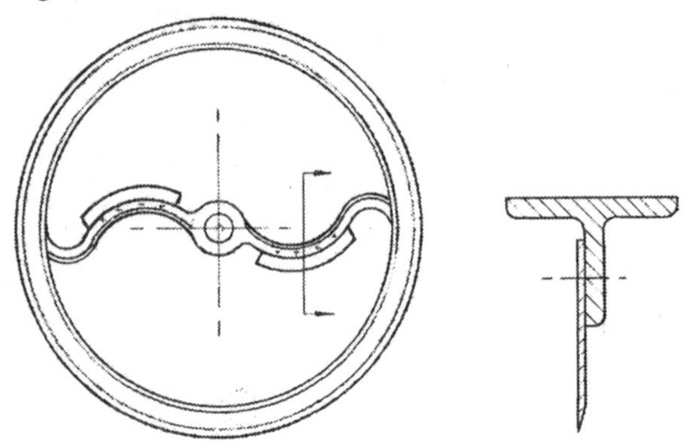

Blades on a chaff cutting wheel.

On wheels made in wrought iron where the parts of the wheel were attached by welding the effect of contraction did not apply. However, although the spokes could then be made radial, the nicer appearance of the curved arrangement continued. Such following of tradition was applied to the fine machines used in clock making where wheels in brass had curved spokes. Obviously, with no technical justification, this was a nostalgic familiarity with a well-known style.

With the advent of machine tools and the need to control slides etc, hand wheels were essential. The hand wheel could be rotated quickly – to save time. For fine adjustment care was necessary and the craftsman would wrap both hands around the hand wheel then, by using a firm grip, a fine degree of rotation was achieved. On a wheel of three inches in diameter an adjustment of .004 inches requires a rotation of about eight degrees.

Hand wheel for machine control.

Ideally, the rim of the wheel should be profiled to be comfortable and textured to ensure a firm grip, especially when the hands are slippery and contaminated with oil. Unfortunately the later development of this type of hand wheel sacrificed the ergonomic features to simplicity and cost reduction.

Turning now (no pun intended) to other types of wheels which were an integral part of a mechanism. The waterwheel with paddles around the periphery would be drenched continually. The traditional construction using spokes, as in the cartwheel was not suitable due to the fretting of the joints and consequent failure owing to the conditions under water and having to transmit a heavy load. To overcome this, clasp arms were fitted which embraced the hub.

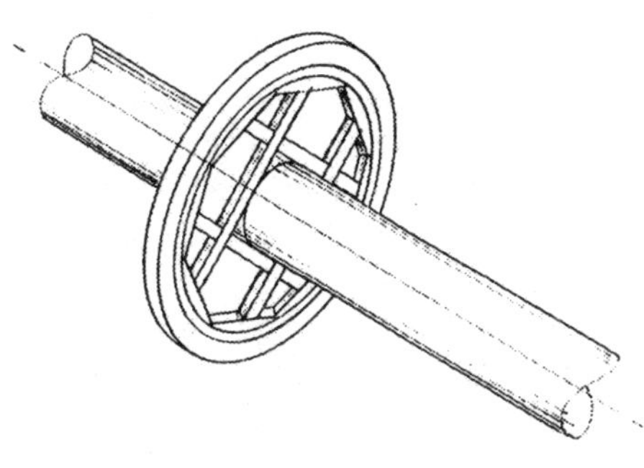

Clasp arms supporting a wheel.

Waterwheels have been used from time immemorial and have either one of two opposing functions. On the one hand they are driven by a stream, a mill race, to operate the machinery in a corn mill or similar, or they are used as a means of drainage to lift water and driven by the machinery inside the mill, usually a windmill. There is a difference in paddle design depending on the use but in both cases considerations of friction and spillage were important to maximise efficiency.

A church bell is mounted on a substantial block of timber which is attached to a shaft capable of holding the weight – often more than a ton. Alongside the bell on the axle is the wheel carrying the bell rope wrapped around a groove in the periphery. The wheel is required to partially rotate in either direction alternately and to withstand the high torque (twisting force) generated by the weight

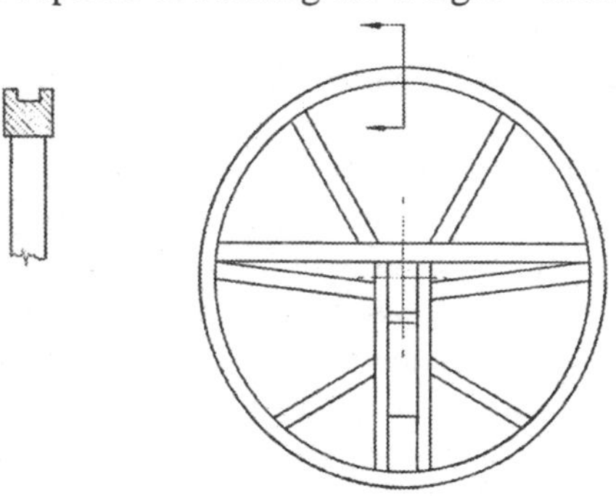

Configuration of spokes on a bell wheel.

of the bell. The support of the rim from the hub is quite different from any other wheel, with the bracing arrangement concentrated on the point at which the bell is mounted. This configuration is unique to the bell rope wheel design for church bells.

Once wheels had been set in motion it was necessary to stop them at some point, especially road wheels on carriages, or the consequences could be disastrous. The friction material attached to the wooden brake block rubbing on the wheel would be leather and the linkage connecting it to the hand lever was often a collection of levers making a tortuous route from wheel to driver. A mechanical advantage was required so that a minimum force on the brake lever would overcome the force of the wheel against the brake block. Band brakes were common on cranes and the necessity for regular inspection and maintenance was vital to prevent an accident should thc leather band suddenly decide to part company.

We are all familiar with the bicycle wheel, where lightweight spokes support a significant load and their success lies in the tensioning, which was first designed in 1808 by George Cayley. He designed a wheel for an aircraft (air borne vehicle) undercarriage and flew a glider in 1809. This was subsequently patented in 1870 by James Starley and William Hillman who were both successful; Starley with the safety bicycle and Hillman who went on to make motor cars.

3:3 Bearings

A revolving shaft in any mechanism must be supported and the housing for such support will be subject to wear. The problem is to reduce the wear which would otherwise increase the clearance between shaft and housing resulting in unsatisfactory operation. The item which has the property of reducing wear is the bearing which rubs on the shaft and fits snugly in the housing.

The environment in which a bearing operates can have a profound effect on the rate of wear. In the china clay industry the trucks carrying the waste material were constantly in contact with the abrasive clay slurry and it was found that a bearing made of oak, supporting the steel axle holding the wheels, lasted much longer than the metal alternative. Curiously, a soft material rubbing against a hard material reduces the rate of wear. A hardened steel shaft revolving in a softer non-ferrous (brass etc) bearing is appropriate and over a period of time the bearing would be intact and the hard shaft worn down. The reason for this is that initially the rubbing action of rotation causes minute particles from the hard shaft to become embedded in the softer bearing which then creates a surface similar to a grinding wheel or sandpaper.

Sometimes bearings are made in two halves – the Pillow Block – especially on long shafts requiring support at the centre or where high loads are encountered. This type of bearing has the advantage that it can be dismantled and the bearing surface renewed when worn out.

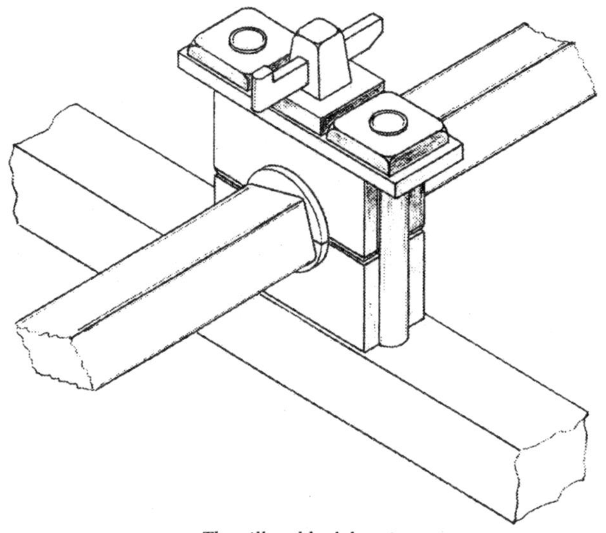

The pillow block bearing.

On windmills and watermills the main vertical shaft rests in an adjustable floating bearing. This was capable of being moved laterally using screws to adjust the engagement of the driving gear teeth. The end of the shaft rested on a simple hemispherical cup. The wind shaft of a windmill was a very heavy baulk of timber carrying the sails and transmitting power to the main vertical shaft of the mill. A wrought iron collar shrunk around the shaft ran in a cast iron ring and the tremendous thrust at the end was absorbed by a cast iron plate. In earlier times such end bearings would have been made of wood or stone. Early

American windmills had a maple wood bearing soaked in oil which lasted longer than the iron shaft which ran in it.

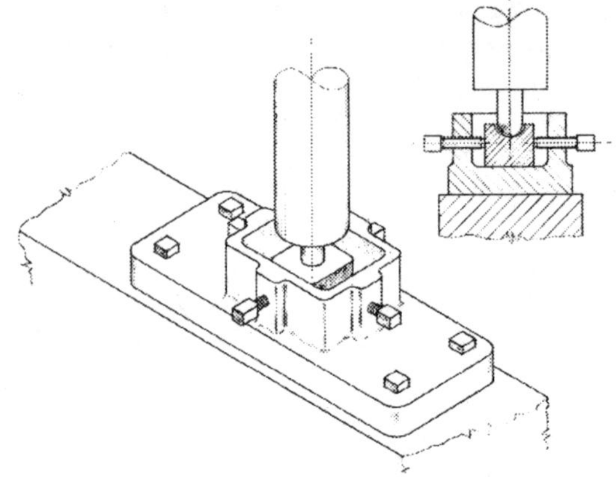

End bearing for a vertical shaft, showing adjustment.

In the 13th century wooden axles on automata and water clocks were fitted with iron 'acorns' nailed to the ends and these were located in iron journals, providing a bearing and preventing side and end movement. A range of bearings was devised by Leonardo da Vinci, but it was not until the late 19th century that these ideas were adopted when precision machining techniques had advanced sufficiently to enable them to be produced accurately.

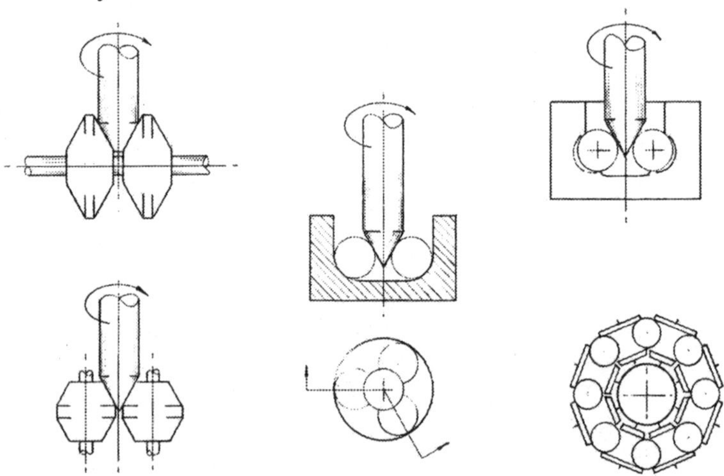

Ideas for bearings by Leonardo da Vinci 1452-1519.

The modern ball bearing (ball race) was perfected in the mid 19th century when the ability to produce truly spherical hardened steel balls was possible. Further development was made when a machine for grinding hard steel balls was produced in Coventry when the bicycle industry was getting under way in the 1880s.

The early steam locomotives were fitted with bearings of bronze lined with white metal. The load on such bearings was extremely high and regular repair would be necessary to maintain the required fit. Such repair work was relatively quick since the bearings were half-bearings. The repair consisted of cleaning out the old worn material, clamping together and pouring in fresh molten white metal. In 1839, Isaac Babbitt invented a tin alloy which is anti-attrition, adapts to a surface, takes a high degree of polish, resists chemical attack, has a low melting point and will not scratch or heat a revolving shaft. This material named Babbitt Metal became the universally accepted metal for lined bearings.

In the early 17th century Robert Hooke recognised that shafts of hardened steel running in bronze bearings were better if dirt and dust was excluded and a constant supply of oil applied. If cast iron is used as a bearing the graphite in the metal provides the means of reducing friction and prevents wear so no lubrication is necessary.

Lubrication of bearings is done by maintaining a film of oil or grease between the mating parts. The speed of rotation is relative to the pressure on the bearing surface. If a thin oil is used for a high speed application it will be squeezed out. Too heavy an oil on high speeds would require additional power to overcome the fluid friction. So, low pressure equals low viscosity and vice versa.

In clock making there must be adequate clearance between the ends of the steel shafts (pinions) and the holes in the brass end plates. Traditionally olive oil was used as a lubricant as it was easy to acquire but although the wear would be reduced, the oil solidified and created a hard film akin to varnish which filled the clearance between pinion and bearing and simply 'gummed up the works'. Gumming is caused by the absorption of oxygen in animal and vegetable oils but mineral oils tend not to suffer from this problem. Another disadvantage of animal and vegetable oils is that they will decay by attracting bacteria and mould. The resultant aroma, combined with a rise in temperature, is not pleasant!

Steel pinion running in a brass end plate on a clock movement.

On a beam engine built by James Watt the main cylinder was apparently lubricated with a mixture of tallow and olive oil to ease the passage of the piston. This lubricant was contained in a bowl shaped depression on the outside of the cylinder which, being warm, kept the mixture in a fluid state. By using a ladle the operator was able to feed the oil through an aperture in the cylinder wall directly onto the piston. It was found at that time that hogs' lard tended to thicken due to the increase in temperature as a result of the friction and linseed oil dried out and created friction.

Lubrication of machines tended to be neglected, maximising production was more important than preventative maintenance. The task of oiling and greasing was given to the sloppiest workers with the result that machines eventually lost their efficiency. By changing the type of lubrication it was possible to increase the speed and power of machines in textile mills. In cold weather a machine which had been standing over the weekend required more power to warm it up on a Monday morning.

Instead of relying on maintenance to lubricate steam engines two solutions were proposed to be applied during manufacture; to use pasteboard soaked in linseed oil or rape seed oil, dried and fixed to the piston, allowing it to wipe along the cylinder, or to surround the piston with quicksilver (mercury) or any metal that will be fluid in the heat of boiling water. Whether either of these suggestions were applied or how successful they were is not recorded.

By the mid 19th century serious efforts were made to find the ideal lubricant. Young in Derbyshire extracted paraffin from coal. Gesner, in America, produced paraffin oil and called it Kerosene. Also, the design of mechanisms was improved to ensure that lubricants stayed where they were supposed to be. This included applying a saturated rubbing pad, cutting labyrinth grooves, drip feeds and grease cups etc.

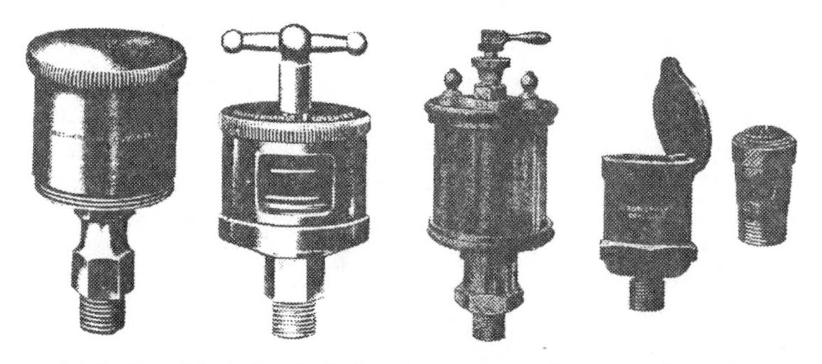

A selection of devices to feed oil and grease to rotating parts of a machine.

With the growth of the machine tool industry it was essential that slideways to guide the cutting tools were absolutely flat, otherwise the products made would not be accurate. Joseph Whitworth, in the early years of the 19th century, created perfectly flat surfaces. He discovered that when two such surfaces are placed together they will adhere (wring). This is not due to atmospheric pressure but, as was discovered many years later, to molecular cohesion. Whitworth overcame the problem by scraping shallow depressions in the surfaces which, when filled with oil, provided a smooth sliding action. The technique of scraping is to use a tool shaped like a flat file (an old file was sometimes adapted) which was sharpened on the end. In use this was pushed into the surface and it removed a fine sliver of metal about one eighth of an inch square and a few thousandths of an inch thick. A master plate, absolutely flat, was smeared with an ink known as engineers blue and rubbed on the workpiece to show the high points to be scraped off. It was a time consuming job even when the surface was nearly flat on commencement. The resultant surface was a pattern of squares which reflected light and created a most attractive design, known as Swiss mottling. An alternative method to create a flat surface was to apply emery powder mixed with oil or water and then rub the two surfaces against each other so that any high points would be worn away. This left a smooth surface but the removal of excess emery was difficult and this residue would cause wear in use. Also, there were no pockets to hold lubrication. Thus, slideways on machine tools moved smoothly as a result of scraping and the problem known as stiction was overcome. Stiction occurs when a slide has to be moved a very small amount and the resistance to motion, combined with the inertia, requires a greater force than that normally applied, resulting in the slide moving more than anticipated.

Metal surfaces rubbing together gave mechanical understanding to the Greek word Tribein, meaning to rub, and so the business of lubrication became the science now known as Tribology.

3:4 Joining and Fastening

Fastening materials together by mechanical means, whether by wood screw, nuts and bolts, nails or any of the various adaptations is something with which everyone is now familiar. This current situation has taken many centuries to develop from the time that body armour was secured by a projecting peg with a cross hole to accept a taper pin.

For many years the pillars separating the main supporting plates for the movement in clocks had the end projections secured using a pin or a wedge.

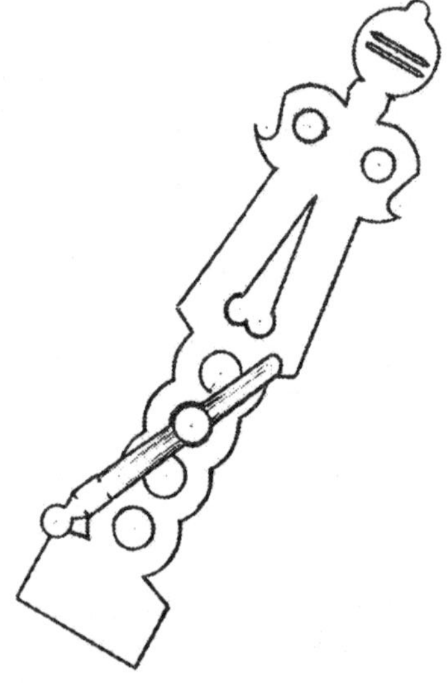

Pin securing plate on a leather jerkin, 17th century.

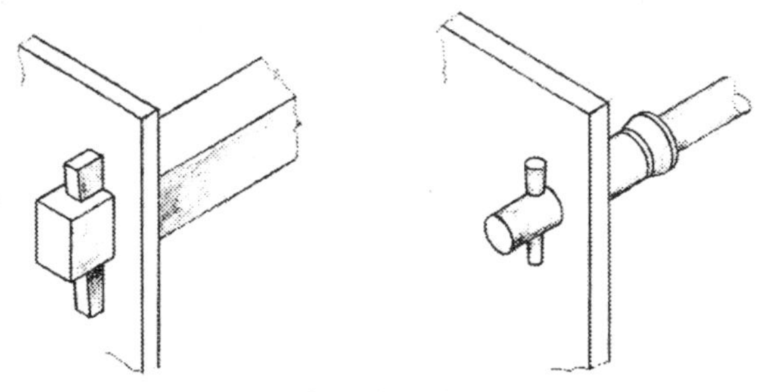

Securing methods for pillars on a clock.

Until the 16th century all door hinges were attached to the face of a door, they were long straps to spread the load and highly decorative with scroll work terminating in natural forms or symbols, fixed at a number of points using large headed nails which were clenched tight. The advantage of covering the majority of the face of the wooden door with ironwork was to discourage house breakers. Doors on cathedrals are particularly ornate and those on parish churches only marginally less so. Butt hinges, which we use today, could not be fitted to the edge of the door, clenching was not possible. They would work loose if attached with nails and the wood screw was not available.

Cathedral door showing medieval attachment of hinge.

After the mid 16th century screws appeared with coarse and irregular threads cut by hand with a nut made to suit. Interchangeability was not possible. The heads of the screws were either made square or slotted for use with a screwdriver.

In 1760 John and William Wyatt patented a method for making wood screws. The operation consisted initially of taking a pre-formed forged iron blank and finishing the conical (countersunk) head using a file on a lathe. This was followed by a saw cutting the slot and the thread (worm) cut using a tool guided by the machine lead-screw. These first woodscrews, i.e. metal screws for fixing wood, had blunt ends and parallel threads, it was not possible at that time to turn tapers on the lathe. Such screws could only be used to join boards together with a nut and could not be used for 'blind' fixing. To cut a point on these early screws would defeat the object since although penetration of the wood was easier there was no engagement with the thread. It was not until the mid 19th century that wood screws were produced with a thread tapering to a sharp point and this happened in America. Screw makers were called Girders, and instead of taking several minutes to make by hand, a screw could be made in six seconds at the Wyatt Brothers' factory. This factory was situated north of Birmingham.

By 1765 there were available, apparently, more than 30 sizes of woodscrew which sold from one shilling and sixpence to 36 shillings per gross. At that time wood screws were used only where absolutely necessary, or by people who could afford to pay the price. Until the end of the 18th century all construction work, even on good quality houses, was fixed using metal nails and wooden dowels. Cabinet making did not involve fastening using metal as wood joints did the job.

It took many centuries for someone to realise that the helix used to generate the force in an olive press could also be made as a threaded nail. In metalwork the screw (bolt), having parallel threads, was made using a die, which consisted of a nut with cutting edges and the nut was made with a tap, which is a screw having cutting edges. This was, and is, the accepted way that steel screws and nuts are made by hand but the taps and dies had to be made accurately on a lathe initially and this work is credited to Joseph Clement, an employee of Henry Maudslay before going into business on his own account. Henry Maudslay, in the early 19th century, was at the forefront of engineering.

The use of screws and nuts as a means of joining metal parts grew, and Maudslay recognised a need for standardisation. He was dedicated to the importance of uniformity and interchangeability and studied the relationship between the number of threads and the screw diameter. For example, fine threads on a large diameter screw could easily be stripped when a load was applied on tightening, thereby rendering it useless. Standardisation of sizes made life easier during manufacture, when any nut of a particular size would fit the same size screw, also when a machine was dismantled for repair or maintenance the screws and nuts which became mixed up were easily identified.

Joseph Whitworth, who worked for Maudslay, experimented with various combinations of screw sizes and numbers of threads. He selected the best and this became the universally accepted British Standard Whitworth (BSW) thread system which survived throughout the 20th century until metrication.

A machine to make screws and hexagon bolts was patented by Whitworth in 1834 although previously Clement and Maudslay had made a bolt threading lathe. Whitworth's design was a

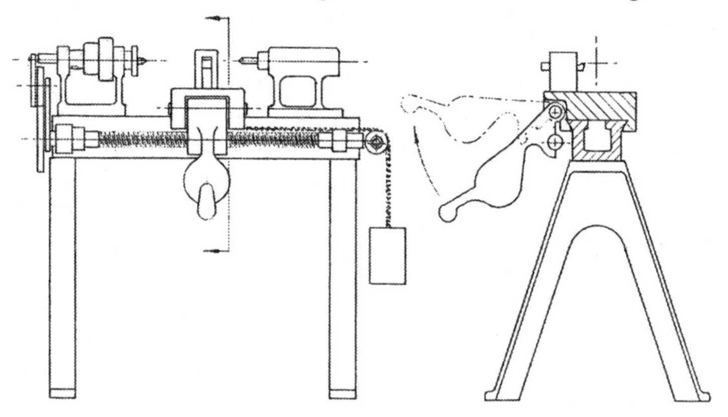

Bolt threading lathe.

significant improvement because it was semi-automatic with a three jaw self centring chuck to hold the workpiece and a die box which, in two passes, could finish cut a screw. The operator only had to load the blank material into the chuck and the machine did the rest. It is interesting to note that the three jaw chuck was ideal for holding hexagonal material, suitable for bolts. This was the technique used in Manchester, where Whitworth operated, whereas in London the bolt heads in the 1830s were of square form.

At the beginning of the 19th century hand made nails were replaced by those made by machine. One method was to stamp them out of sheets of wrought iron, or alternatively, to use the slitting mill, which was introduced from Flanders in 1590, where iron bars were passed through successive rolls to reduce the thickness to the sheet size equivalent to the nail and then fed through a pair of rollers with cutting collars. The resulting length of rod, either square or rectangular in section, was cut to nail length and finally the point sharpened to a chisel edge.

Round nails, when driven across the grain, tend to force the fibres of the wood apart causing splitting. This is overcome by first flattening the nail point with a hammer and in effect making it blunt; a cutting action is then exerted on the fibres rather than pushing them aside. Such a problem is not normally encountered with rectangular section nails. Nails rely on friction to grip timber together and a number of them in a row used to join two boards together should not be driven in perpendicular to the wood but alternately at an angle, skew or dovetail nailing, to avoid them pulling out.

A nail is best suited to joining a thin piece of wood to a thicker one, it will not hold firmly in thin material. A wood screw, on the other hand, is effective however short, and applications included the fixing of the leather sides to a bellows board and attaching a matchlock to a gunstock where the screw could not be allowed to penetrate inside and interfere with the mechanism.

There is an anecdote which claims that Alexander Nasmyth (father of James), around the year 1810, needed to repair a stove using rivets. He did not wish to disturb his neighbours with the noise of hammering rivets so he heated the rivets and pressed them into place. This system of driving hot rivets was long afterwards patented and used in the construction of bridges, shipbuilding and boiler making where, apart from the percussive noise reduction, the effectiveness of a joint is better than using cold rivets. Riveting was a skilled job demanding three people working in perfect harmony. The rivet was heated to red hot then passed to the person who popped it into the aligned holes on two overlapping plates, placing the 'dolly' over the head of the rivet and the riveter on the other side clenched the protruding end to create a rounded head. This operation had to be performed very quickly whilst the rivet was still hot. As the rivet cooled, and contracted, the plates were drawn together making a watertight joint.

Richard Roberts was responsible for inventing a mechanism for punching holes to take the rivets in the construction of the Britannia Bridge over the Menai Straits by Robert Stephenson in 1847. The equipment was supplied as a result of labour problems in punching the holes manually and was capable of 40 holes per minute. Each rectangular steel hollow section beam needed 327,000 holes for the rivets.

In the late 19th century James Watt produced clear directions for the fitting of rivets in boiler construction and recommended that joints above the water level should be closed using a blunt chissel (sic) and wetted with a solution of sal ammoniac in water, or rather in urine, which by rusting and consequently expanding, would make the joints steam tight.

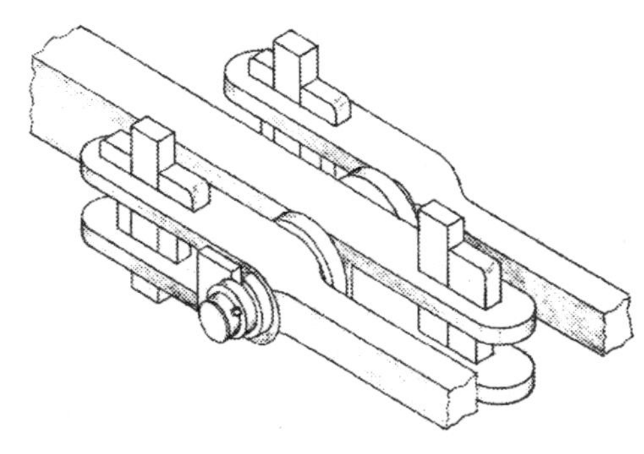

Coupling for pump rod on a mining engine.

The connecting of shafts in line to transmit power creates an interesting design problem, especially when alignment of the shafts is not true and consequently flexibility has to be allowed within the coupling arrangement whilst maintaining effective power transmission. The modern universal joint neatly solves the problem, but in the 18th century no such refinement was possible and although relatively crude, the arrangement worked, as applied to the connections for pumping rods and the piston rod end in early steam engines.

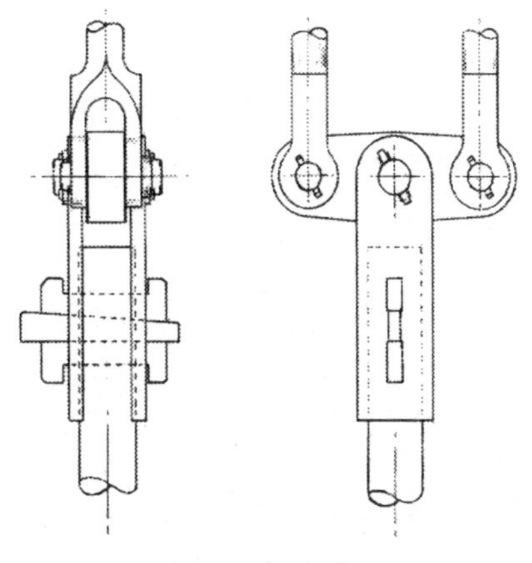

Piston rod coupling.

Substances for joining materials together such as adhesives have existed for a very long time. By trial and error they were found to stick and to do the job adequately, being used by many different craftsmen, but it was only in the 20th century that the mechanism which causes things to stick together was found to be molecular attraction.

Finally, some advice from James Watt, who allegedly made this proclamation in 1816:

> 'I beg to advise your worker to always rub his wood screws on some candle grease before he screws them in; it will tend much to his own ease as well as to that of him who unscrews them.'

A technique still used to this day and every woodworker has a stub of a candle handy and it does make driving a woodscrew much easier.

3:5 Structures

Any arrangement to support a load or to raise an object to a higher level must be safe, substantial, durable and cost effective. In the past, structures were massive and cumbersome, just to be on the safe side, in effect they were 'over engineered'. Only by trial and error and the accumulation of past experience, together with convincing argument (with the patron or client), was progress made. As materials were improved, and a body of knowledge built up based on scientific principles, it was possible to calculate and predict the behaviour of structures in service. Even now, despite the unprecedented developments in technology, a safety factor of 5 or 10 is applied to ensure success.

Ironbridge. Centre of parapet railing.

The most famous of metal structures is Ironbridge over the River Severn in Shropshire, built of cast iron in 1779. Other cast iron bridges were built during the same period but sadly they have not survived. The shape of Ironbridge is a graceful semi-circle with economical lines and it retains its original features today.

The design is based on the available knowledge of the time and although the various members are of cast iron they could at first glance appear to be wooden. The details of the construction are akin to woodworking technique since the builders were familiar with timber construction. No bolts were used in the fixing, only pegs and dowels with shouldered dovetail joints. The ends of the cast iron ribs, where they meet the ground forming a plinth, are decorated and closely resemble the moulding used in cabinet making work. The inclusion of a circle and ogee motif in the structure adds aesthetic appeal and goes some way to offset the harshness of the appearance of cast iron. The architect was Thomas Pritchard who was also a joiner and, therefore, fully conversant with the working of wood. Two traditions combined to produce Ironbridge – the innovation of the ironmaster and the craftsmanship in wood and stone. As a result two professions began to emerge, the architect and the civil engineer.

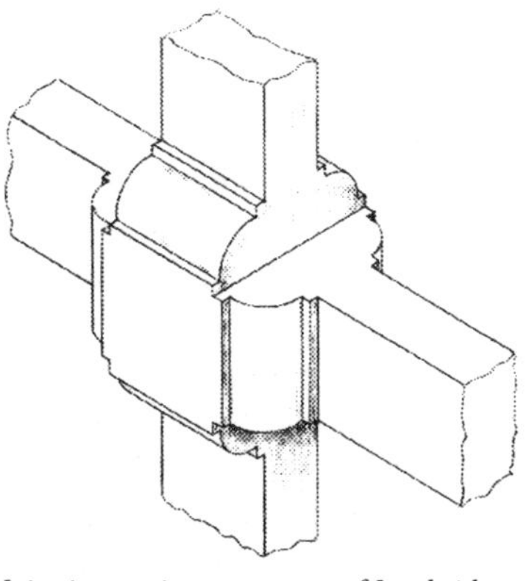

Joint in cast iron structure of Ironbridge.

The ribs of the bridge are the main load bearing elements of the structure and each consists of two quadrants joined at the top to create a semi-circular arch. Each rib weighs nearly 6 tons and was possibly cast on site by the banks of the River Severn rather than at the foundry at Coalbrookdale a mile away. The faces of the ribs are not quite flat and the concave depressions represent local contraction of the iron as it cooled. Also, there are some holes evident, called blow holes, and these are due to the presence of air when the molten metal was poured into the mould. This great symbol of success in the use of cast iron to create a river bridge was due to Abraham Darby III who followed in a dynasty of ironmasters to pioneer the construction of such massive components.

Ironbridge having a truly semi-circular arch presumably made sense at the time, using cast iron rather than copying the semi-elliptical shape of the stone bridge tradition. It was realised much later that a catenary curve turned upside down was a more appropriate shape and a wider span was possible. The catenary curve is generated when a uniform section flexible cord is suspended from fixed points at the ends, similar to a washing line without the washing! It was Robert Hooke who, in the 17th century, defined the catenary and recognised that its inversion created a rigid arch; but long before this there is evidence in churches, where the catenary is contained within the stonework. This would indicate that the medieval stone mason knew exactly what was needed but was possibly unaware of the mathematical reason for it.

Cast iron pillars and beams appeared for the first time in architecture in 1796 at the flax factory in Shrewsbury built by A. W. Skempton. The use of cast iron created more space for the looms than was previously possible with timber construction. There is the suggestion that the risk of fire was reduced although the floor boards were still wooden. Pillars in cast iron are evident today in many old buildings and being indoors have survived in excellent condition, they are highly ornate with fluted columns and decorated capitals with moulded bases, showing a strong classical influence. The substantial supports for massive beam

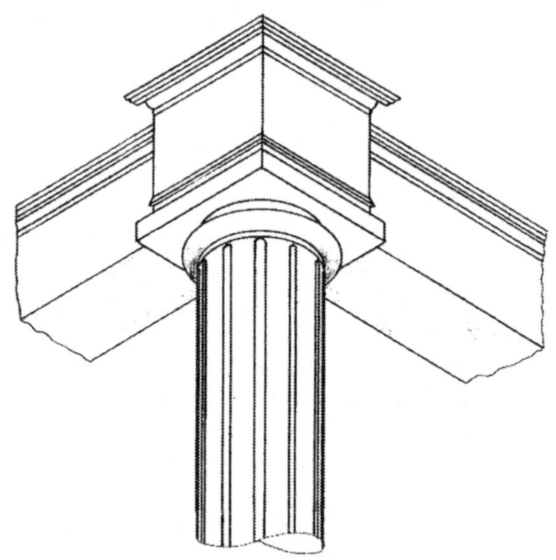

Cast iron column supporting wooden ceiling beams.

engines are similarly ornate with the features picked out in coloured paint. These look somewhat out of place in the environment of a noisy lumbering machine but we can appreciate the aesthetic quality despite the distraction. Beam engines require access for inspection and maintenance purposes at various levels. The platforms high in the air have safety railings and the stairways have banisters. Both these features were cast with ornate designs giving an incongruous aesthetic appearance in such austere circumstances. A cheaper and more utilitarian design for the railings and banisters would have been appropriate, but it says much about the need to make things look nice.

Cast iron balustrade.

Tubular cast iron was first used in 1803 for a bridge in Suffolk. The use of cast iron gathered momentum and included using it in church buildings to replace the traditional masonry in the ornamental details as well as the structural members.

The spires surmounting church towers built some 600 years ago did not always rely on the quality of the mortared joints in the stonework or the internal timber structure to maintain stability. The simple, yet effective, method was to suspend a heavy mass of iron from the tip of the spire and allow it to hang down well within the tower. This had the effect of lowering the centre of gravity of the spire and would permit deflection under wind pressure without structural failure or collapse.

A beam always needs to be supported and if such supports are placed at either end the beam will tend to sag in the centre, conversely if they are close together the ends of the beam will droop. It was established by Sir George Airy that the optimum distance between the supports (span) should be 0.577 times the beam length, i.e. just over half the beam. In practice it appears that the supports are too close to the ends but in fact this does avoid any deflection in a uniformly loaded beam.

The Crystal Palace, built to house the Great Exhibition in 1851, was designed by Joseph Paxton. The building was light, airy and graceful, it was also large. The roof design was inspired by the Royal Water Lily plant which contains a complex series of ribs to spread the six feet diameter leaf over the surface of the water.

St. Pancras Station in London was built in 1867 and for over a century was the only building in the world having a roof span without a central support.

In Roman times, lead was used to secure the joints between blocks of masonry by filling grooves in adjacent blocks. An alternative was to use an iron stirrup or staple, the ends being embedded in lead which was poured in to secure the iron to the stone. Another example was in securing the drive shaft to the runner stone in a corn mill.

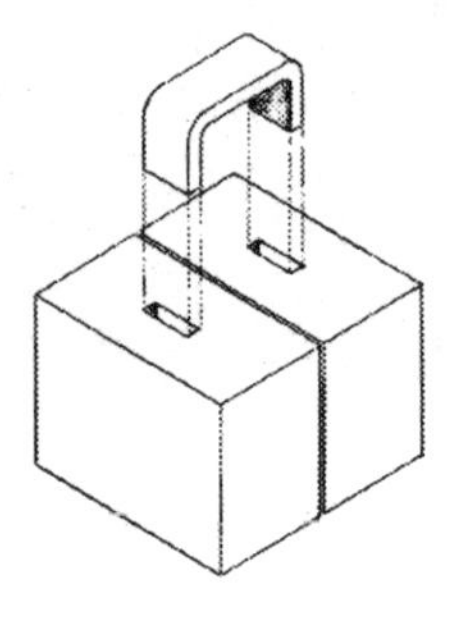
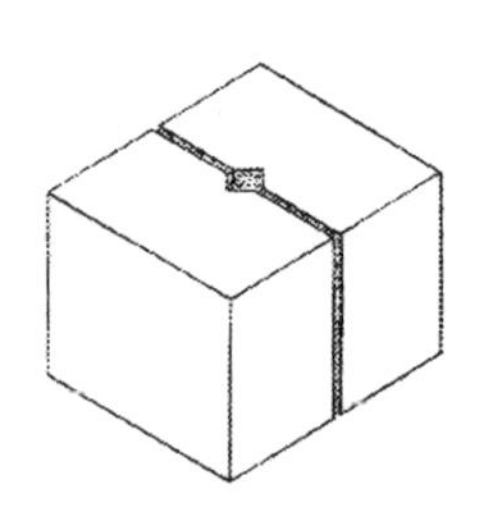

Methods of securing masonry blocks.

These examples of attaching iron to stone using lead require care in practice. If the iron or the stone is damp when the molten lead is poured in at a temperature of 330 degrees Celsius there is a tendency for the lead to spit and possibly cause injury. To overcome this, the hole was initially filled with wax which melted as the lead was poured in. There is an affinity between lead and iron which allows intimate contact and so ensures security.

Fastening large stone blocks together using iron staples continued into the 18th century and unfortunately, in some cases, the weather would penetrate the smallest gap in the stonework and attack the iron. This corrosive action (rust) caused the iron to expand and resulted in cracks in the stonework and ultimately collapse. This situation is particularly evident at the Mausoleum in the grounds of Castle Howard in Yorkshire, where it would appear that lack of maintenance and neglect has resulted in dereliction of an outstanding building. At the Parthenon in Greece the same technique of stapling stonework was used, but either the metal was corrosion resistant or the climate more favourable as there is no evidence of cracked stonework in this ancient structure.

Steel rods are now used to reinforce long concrete beams. The temperature change due to climate alters the length of the concrete beam significantly, but since the co-efficient of thermal expansion is the same for concrete and steel intimate contact is maintained.

Ancient devices for lifting heavy weights show ingenuity in simple design. The Lewis fits into a dovetail shaped mortise cut into the top of a stone block which could weigh up to half a ton. As the block was raised so the Lewis would bite into the dovetail aperture and the greater the load the more the Lewis gripped. Another device was based on the scissor principle and, once

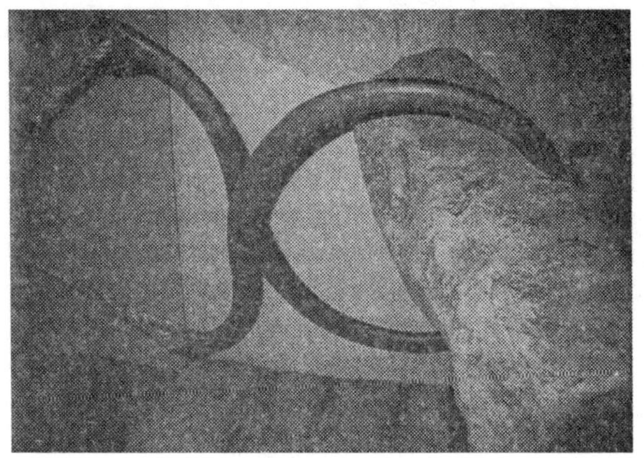

Scissors lifting device.

again, the grip on the stone increased with the load. In moving millstones it was necessary to manipulate them for dressing and the lifting arms were inserted into sockets.

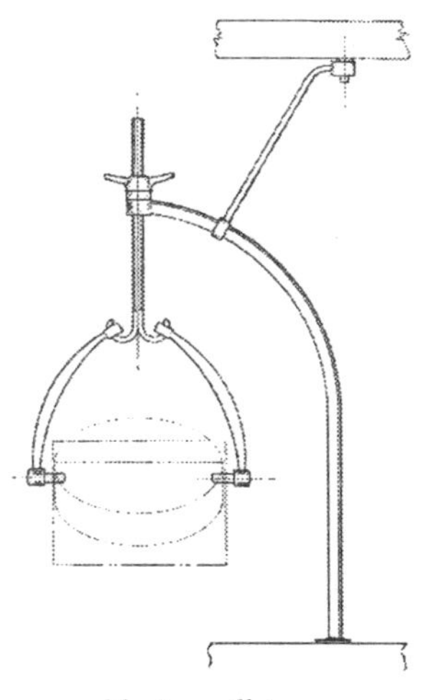

Moving millstones.

When a thin sheet of metal projects from a wall as an awning, etc., there will be a tendency for the outer edge to droop and this is overcome by providing corrugations along the sheet perpendicular to the wall which create stiffness. This principle also works on thin strips of metal by placing a vee shaped indentation along the length. There is a parallel in the natural world where long leaves on plants are so shaped to prevent drooping.

An opinion was expressed by James Nasmyth, in the mid 19th century, who travelled widely in France and took the opportunity to study closely the architecture both ancient and modern. He commented thus:

> 'It appears to me that one of the chief causes of the inferiority and defects of Modern Architecture is that our designers are so anxious to display their taste in ornamentation. They first design the exterior, and then fit the interiors of their buildings into it. The purpose of the building is thus regarded as a secondary consideration. In short, they utilise ornament instead of ornamenting utility – a total inversion, as it appears to me, of the fundamental principle which ought to govern all classes of architectural structures. This is, unfortunately, too evident in most of our public buildings.'

In retrospect we can recognise the validity of these comments and either admire or criticise the exuberant ostentation of Victorian architecture.

4
Mechanisms

A machine is a collection of mechanisms working together to make it operate correctly, whether a bicycle or a sewing machine. Also, a mechanism can be simple and independent, such as a door hinge. In all cases mechanisms are essentially moving parts and to work well they must be reliable. To this end they are accurately made of appropriate materials and maintained throughout their lives.

Our understanding of what mechanisms do has changed little from the perception during the 18th century, as the following extracts from a 1772 dictionary shows:

Machine – A contrivance, or piece of workmanship composed with art, and made use of to produce motion so as to save either time or force. An engine.

Engine – An instrument consisting of a complication of mechanic powers, such as wheels, screws, levers etc. United and conspiring together to effect the same end.

Mechanism – Action according to mechanic laws. The construction of the parts depending on each other in any engine.

Mechanic – A manufacturer; or one engaged in trade or low employments.

The above extracts are a faithful reproduction except for the substitution of the letter 's' instead of 'f'; e.g. the word 'consisting' was written as 'confifting'! The definition of a mechanic is interesting as it reflects the status of the craftsman at the time.

To return to the subject of mechanisms, the designer/craftsman would take particular pride in producing an item which was capable of doing more than one job. This is evident in clock making where space is at a premium and a collection of a number of bent pieces of metal and levers would lead to wear and become unreliable.

After previously measuring time by water clocks, mechanical clocks were developed in the 10th century and apart from changes in the escapement mechanism (the control evidenced by ticking) are little different today, with a weight or a mainspring providing the power via a gear train to drive the hands. Absolute precision in manufacture is necessary for accurate timekeeping but there must be freedom of movement in the gear train and no binding or tightness can be tolerated.

It is interesting to consider that the invention of the means to measure time 'physically' deprived us of the ability to tell the time by natural activity. There are parallel examples in other fields where the ancient ways of doing things have given way to progress. Sometimes it is recognised that the old way wasn't so bad after all and revival is welcomed, and the value of ancient knowledge is confirmed. Evidence in recent times is the revival of the use of herbal remedies.

Turret clock movement.

The oldest surviving clocks in England are at Salisbury, Wells and Cotehele and date from the 14th/15th century. They were made in iron and the hours marked by the striking of a bell; minutes were not shown as there were no hands or dial. The clocks at both Salisbury and Wells cathedrals dating from 1386 are turret clocks, where the mechanism is contained in an iron frame which is decorated with ornate projections. This feature was common at the time and could be regarded as a demonstration of pride by the craftsman since the turret clock would have been tucked away in the cathedral with no opportunity for anyone to admire the work.

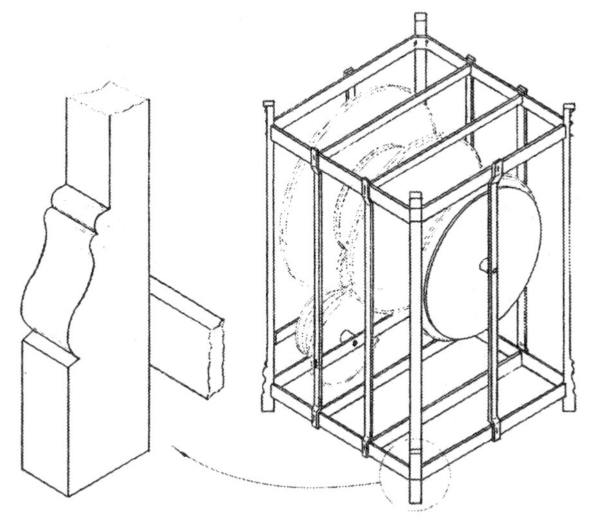

Decorative frame on a turret clock.

The 1490 clock at Cotehele House, in Devon, does not conform to the turret clock construction in that the gear train is laid out in a vertical line mounted on a strip of iron. The teeth on the gears vary in design depending on the work they have to do, those which move quickly with little load are long and thin and heavily undercut to engage smoothly with mating teeth, whereas, the teeth on the gears nearest the weight are larger and of triangular shape, to cope with the higher power.

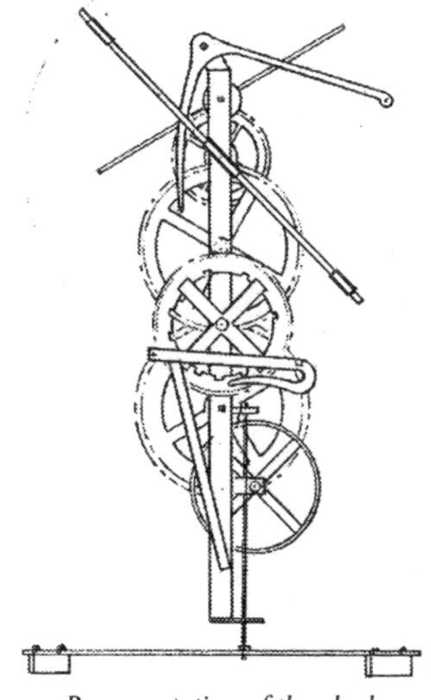

Representation of the clock at Cotehele.

One of the problems with historical data is the sometimes conflicting records of credit for inventions. There were people who were widely dispersed seeking answers to the same problem, since at any one particular point in time a common technological situation will tax the intellectual minds of similar disposition. Those having influence or the right connections will get credit. An example of this is the invention of the miners' safety lamp, where Humphry Davy and George Stephenson both simultaneously worked on the problem of avoiding gas explosions in mines. They both came up with a solution, a lamp, within weeks of each other and the argument as to who was first persists to this day. Davy was given a handsome reward and Stephenson received a relative pittance.

Another possible situation where credit was not entirely acknowledged was in the case of the introduction of the pendulum, when one Richard Harris working in London allegedly made a pendulum clock in 1641. Nevertheless, the most significant improvement in clocks was the discovery in 1581 by Galileo of the time keeping properties of the pendulum, and the application of this by Huygens in 1656. A pendulum of one metre in length will give a one second swing. It has to be freely suspended to allow oscillation from side to side but also to move backwards and forwards, describing a figure of eight path to satisfy natural laws. Superseded in about 1800 was the mechanism known as the verge and foliot, an arrangement with a swinging arm driving two 'paddles' alternately to contact teeth in a crown wheel; a device requiring careful setting up and regular maintenance. During the 17th century the anchor escapement – the time-keeping heart – was developed and with the pendulum regulating the pattern was set for the future.

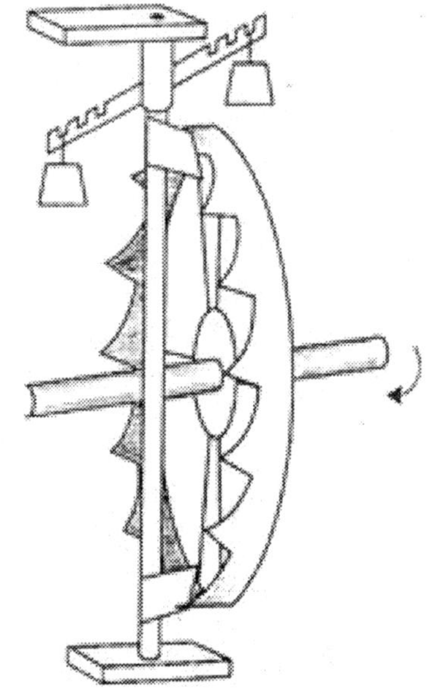
English verge escapement.

Now, to return to the 1772 Dictionary for the contemporary definition of a clock:

> 'A machine going by pendulum, serving to measure time and show the hour by striking on a bell.'

Notice there is no mention of a dial or hands.

The alternative to the pendulum to control time was the spring driven balance and although this was applied to the watch in 1675, it was not sufficiently accurate for use on board a ship. A pendulum needs a clock to be mounted on a firm foundation to work accurately and is, therefore, also unsuitable for use on a ship at sea. The balance would not be affected by external influences but it needed refining and this was done by John Harrison, who spent his entire life perfecting a timepiece capable of withstanding the rigours of a sea voyage, which in turn enabled the calculation of longitude. This was a work of renown and he did successfully produced an accurate marine chronometer.

The escapement – pallet and scape wheel.

The impetus for this work came from the government who offered a reward of £20,000 to anyone who could produce an accurate timepiece. Harrison received small contributions during construction but was denied the full reward due to the authorities making petty objections, and it was only by Royal intervention that he finally received the total amount. This was 45 years after Harrison first started on the work and he only survived three years to enjoy his just reward. He was a woodworker by trade and made a wooden clock when he was 22 years old in 1715. He had an outstanding understanding of the properties of materials and their appropriate use. Oak was used for gear teeth; pivot bearings were of lignum vitae, which is extremely hard and naturally oily; the ends of the pallets had knife edges which rocked on small pieces of glass, virtually friction free. John Harrison also invented the compensating pendulum which counteracted the effect of temperature change causing expansion and contraction resulting in inconsistent time keeping. He noticed that different metals had different rates of thermal expansion, so he made a frame like a gridiron with alternate rods of brass and steel arranged so that those which expanded the most were compensated by those which expanded the least.

To appreciate the fine design of the balance wheel and hairspring as we know it today it must be realised that it has to work at an extremely rapid speed, 24 hours a day continuously, without attention for many years.

The mechanism to determine the correct striking of the hours was very important in the early days as, without hands, this was the only way of telling the time. Initially this was the count ring,

Striking clock movement. Count wheel system.

a wheel with notches on the outside edge which were progressively spaced at increasing distances apart. A lever riding on the wheel would allow the appropriate number of blows to be struck until it dropped into a notch to stop the striking. This method was superseded in 1670 by the rack striking mechanism

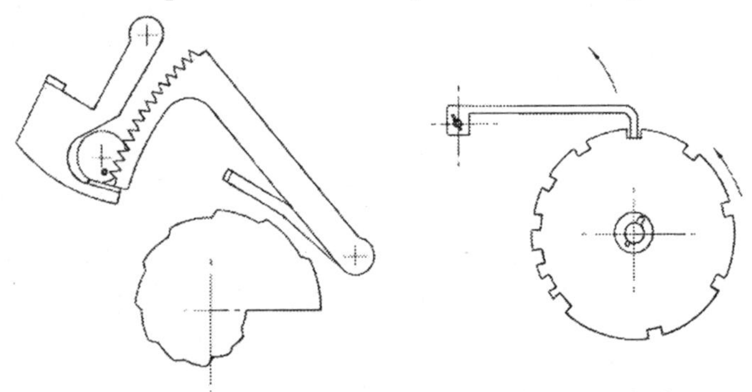

Control of striking on a clock. By a rack and a count wheel.

which gave a more precise way of controlling the striking, and the number of strikes was governed by the rack lever falling onto a snail cam.

Clocks were originally driven by weights which delivered a constant force but this system was superseded by the spring. On spring driven clocks, when the spring is fully wound it will produce a large force and as it runs down this force gradually reduces. The effect of this is that a clock will progressively lose time as the spring loses power. To overcome this problem and maintain a consistent time the fusee is used which first appeared in the 15th century. This item is cone-shaped and has a spiral groove around which a cord is wrapped and connected to the barrel containing the drive spring. A constant force is thus maintained to the movement, since the reduction in force by the spring unwinding is compensated for by the increase in diameter of the fusee. In other words, a consistent torque is the power from the spring multiplied by the radius of the fusee at the point where the cord contacts

The fusee.

Making a fusee in the 15th century must have been a real challenge given the simple lathes available at the time. A great deal of skill was required to cut, by hand, the groove, like a screw thread, on a rotating metal cone. The cord connecting the spring barrel to the fusee had to be extremely strong and gut (an animal fibre) was found to be suitable, but it tended to stretch or contract according to the ambient temperature and humidity. Chains made of steel did not have such a disadvantage and these replaced the fusee gut cord in the mid 17th century. The chain was identical in design to our now familiar bicycle chain, with either three or five rows of links depending on the strength needed. Fusee chains were used in watches as well as clocks and the very smallest could pass through the eye of a sewing needle. When it is considered that thousands of identical links about 0.030 inch long were made and assembled with minute pins, the degree of skill required was quite remarkable.

The fusee chain making industry was centred in Hampshire, around Christchurch, where hundreds of women and children, some as young as nine years, were employed. Their eyesight suffered. It is interesting to note that despite the advances in plastic materials there is apparently no cord currently available to compare with gut or chain to transmit the high force generated by a clock mainspring.

Following the advances in clock design the American industry made its presence felt and Eli Terry in Connecticut made mass produced wooden clocks in 1807, and shortly afterwards invented the 30 day movement. He progressed to movements made in brass and it was claimed that three men were able to cut all the gears for 500 clocks in one day.

The beam of a beam engine is pivoted at its centre and each end swings in an arc. One end is driven by the piston of the steam engine and the other end does the work. In James Watt's engines the piston rod moving in a straight line had to be firmly attached to the end of the beam which moved in a circular arc; such an arrangement would cause strain and the solution to this problem was to incorporate the straight line mechanism.

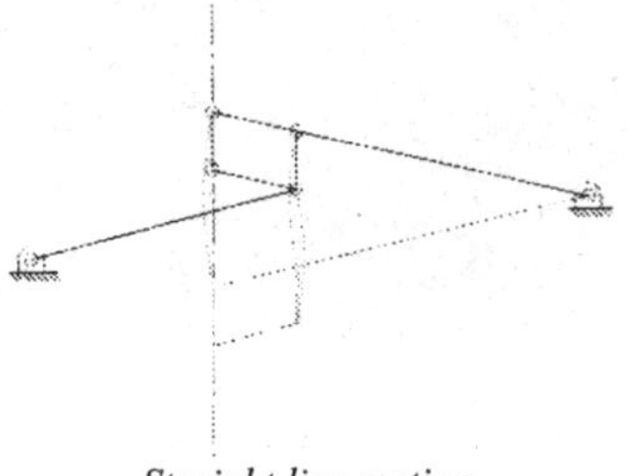
Straight line motion.

This is just one example demonstrating the intellect of the early engineers, where they were able to solve the most difficult problems by drawing on their knowledge of mathematics and physics.

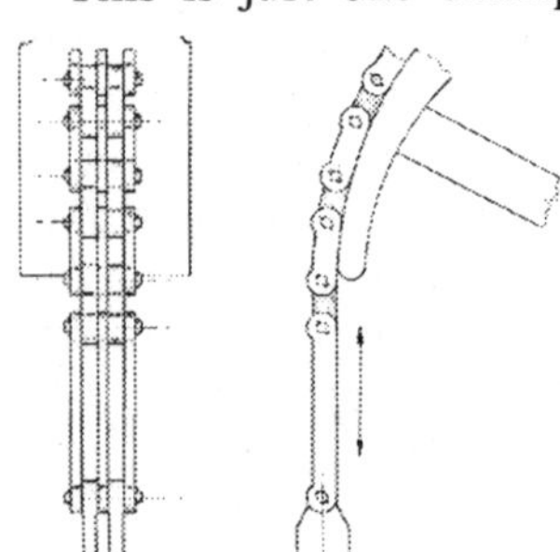
Piston rod to beam connection.
Newcomen atmospheric engine.

On the earlier Newcomen atmospheric beam engines the piston was only powered on the 'in' stroke and the straight line mechanism was not necessary, the connection made simply by a chain.

The system of control which we now know as negative feedback was used in a number of situations in the past. The gap between millstones grinding corn needed to be adjusted depending on the speed of the driving force, whether by a waterwheel in a watermill or the sails of a windmill. This was achieved by the tentering mechanism which responded to the change in speed by reacting to the centrifugal or pendulum governor and automatically adjusted the millstone gap.

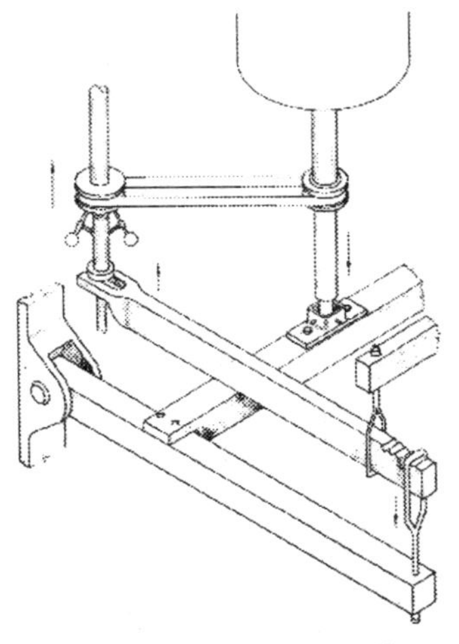

Tentering mechanism on a mill. As the speed of the millstone increased so the pendulum governor tilted the various levers to increase the gap between the stones.

At the end of the 18th century James Watt needed to control and maintain the speed of his steam engines and used the centrifugal governor. This worked by weights moving out from a rotating shaft and, by means of levers and linkages connected to a butterfly valve, it reduced the steam inlet as speed increased.

Centrifugal governor on a traction engine.

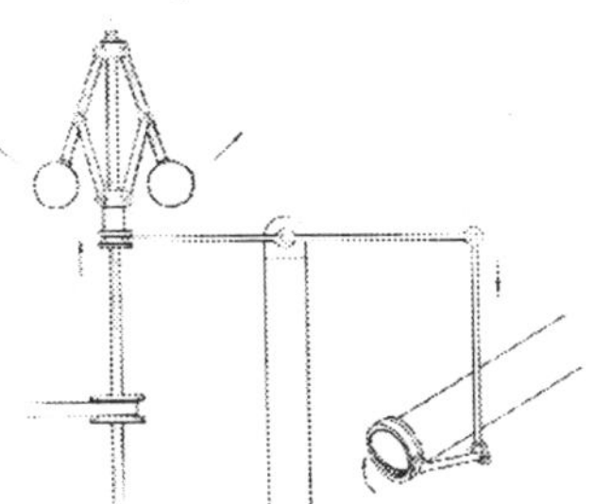

Principle of the steam governor. As the speed increases so the balls on the governor fly out and raise a lever which is connected to the butterfly valve and this then closes to reduce the steam flow and slow down the engine speed.

An early example of what we now know as the servo-mechanism is the luffing of a windmill. Luffing (bringing the sails into the wind) is done by turning the cap of the windmill automatically and this avoided the necessity of the miller doing

e

the job manually using a windlass. The system involved fitting a fantail (a miniature set of sails) to the cap on the opposite side from the main sails, and this would operate like the blade of a weather cock and turn the cap through gearing on the windmill, bringing the sails into the wind.

One simple, yet effective, mechanism is the windlass which has been in use since ancient times. An operating lever on a wheel of ten feet diameter attached to a winding shaft of one foot diameter gives an advantage of 10:1, and by using a simple pulley system this is easily increased to 20:1. So, one man exerting one hundredweight of effort can lift a stone weighing one ton.

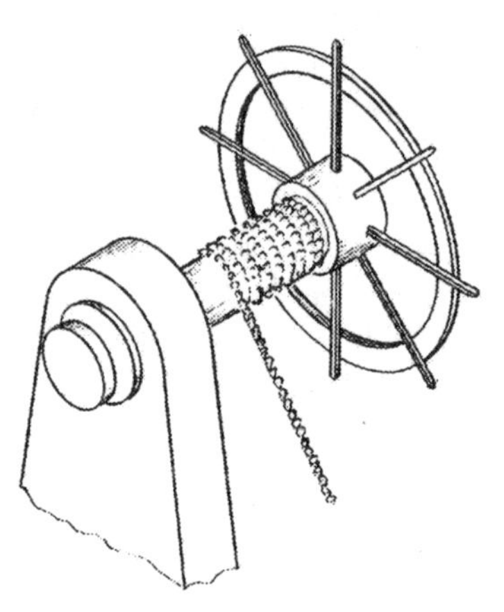

The windlass.

This helps us to understand how it was possible to build mighty cathedrals without the equipment familiar to us today.

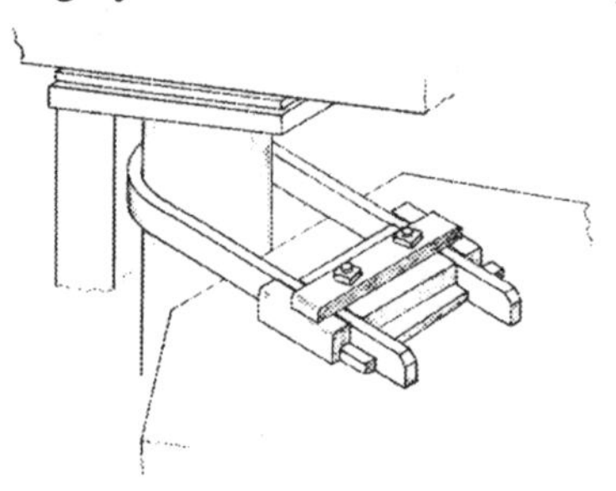

Canal lock gate quoin.

Another simple device is the means of both locating and sealing the lock gates on our inland waterways. The hinge for the gate is a very loose fitting collar which allows the heel post to fit snugly against the stone buttress – the quoin – and the greater the water pressure the tighter the seal.

In the days when joints of meat were roasted on a spit in front of an open fire the smoke jack was used to ensure that every part of the joint was presented to the heat for cooking. The power source to the spit was supplied by a fan in the chimney which rotated due to the smoke rising from the fire, and this was connected by a system of shafts and gears to transmit the power to rotate the spit. When the fire burnt fiercely the smoke would increase and cause the fan to rotate faster, so turning the spit more quickly. A more simple design is a clockwork mechanism,

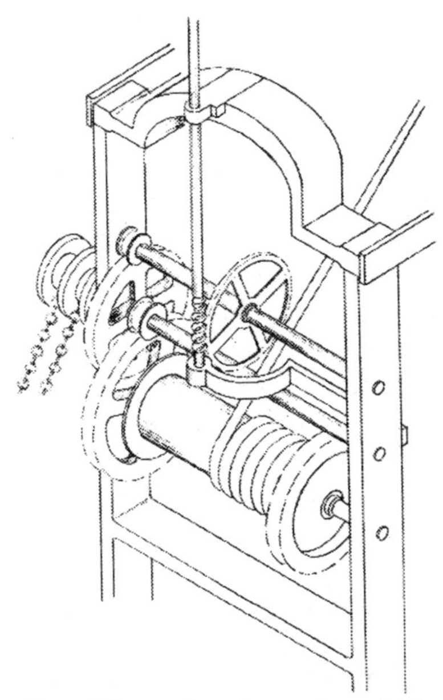
Control for rotation of a roasting spit. The rope at the top right is connected to a weight and drives the mechanism via a reverse worm and wheel, resulting in the pulley at the end rotating slowly and driving the chain attached to the spit.

mounted in a canister, making the hook holding a small piece of meat rotate slowly. Another alternative design to drive a spit was to use a clockwork mechanism driven by a weight with the rate of rotation controlled by an air vane, which consisted of a pair of paddles mounted at either end of a rotating bar, their speed being governed by air resistance.

An item which may loosely be described as a mechanism is the simple water tap. The ability to effectively turn on and off a water or steam supply must have taxed the ingenuity of the early engineers, judging by the various methods they devised to seal the flow. The water tap has been made with a degree of decoration, in some instances completely out of proportion to its function, and we can only admire the time and trouble taken to endow such an object with beauty.

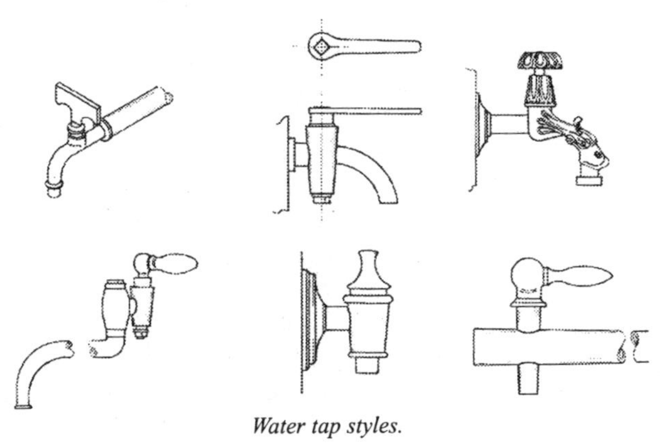
Water tap styles.

One type of mechanism which is often overlooked and taken for granted is the device used to secure doors and gates. There are a variety of different designs either fastened by gravity or spring assisted.

The latch design needed to be easy to operate to ensure a secure closure and the associated metalwork was either simple and plain or embellished to emphasise the status and wealth of the owner, such work must have given the blacksmith an opportunity to show his ability by applying ornate decoration.

Parish church doors have good examples from the basic to the ornate, operated by a cam attached to the door knob and fastened by gravity. Sometimes the lever was decorated with a series of lines and dots. This is thought to represent the Cross of St. Andrew or St. George and is derived from the medieval custom by masons of scribing such marks on the stone doorstep (threshold) of a house to ward off evil spirits. The blacksmith was presumably following this ancient tradition to protect churches.

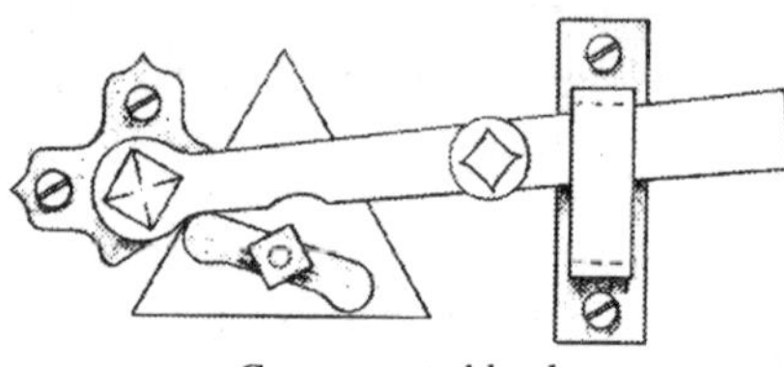

Cam operated latch.

A latch relying on gravity normally rests on the catch with the pivot positioned to achieve this. However, there are examples on gates where this is reversed and the latch is upside down, the handle part being made heavier to keep the latch in the 'up' position. The possible reason for this is that over a period of time a gate may tend to droop and provided that the latch had sufficient movement it would continue to work.

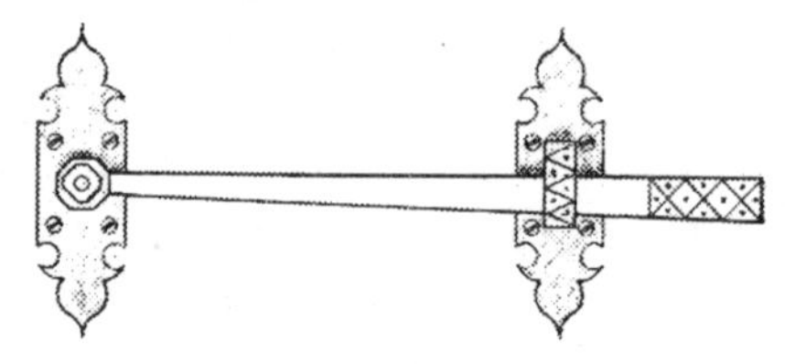

Symbols of saints struck on the end of the latch.

Cupboard door latches can be 'upside down' and assisted in operation by a leaf spring. Being indoors they would not have

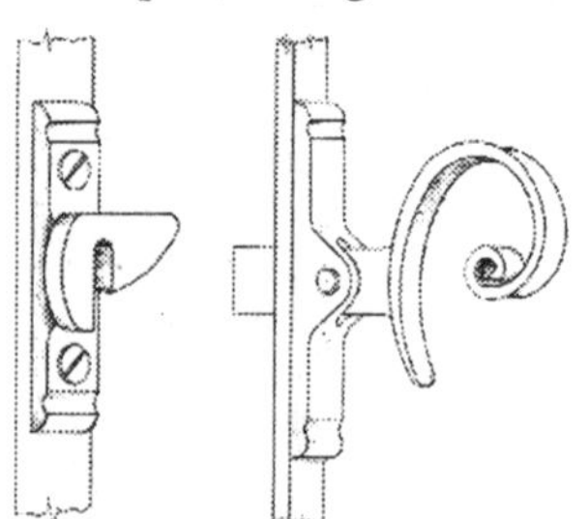

Upside down latch.

suffered the effects of weather and we can now admire such nice samples of brass work.

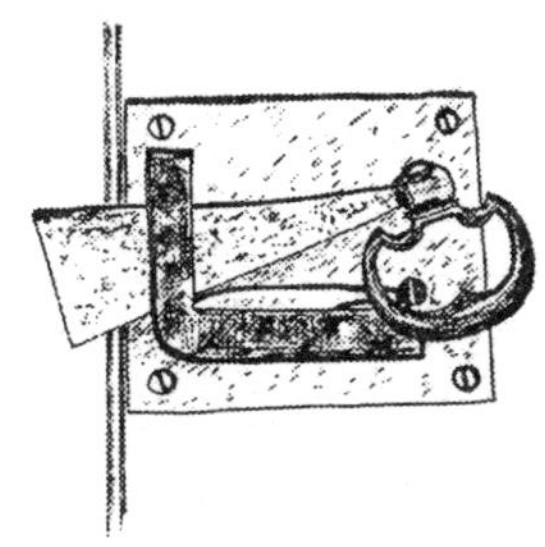

Cupboard door latches. Left: upside down. Right: normal.

Gate catches are normally operated by hand from a standing position, but if you are sitting on a horse this is tricky and on bridle ways an extension to the catch is provided to avoid dismounting.

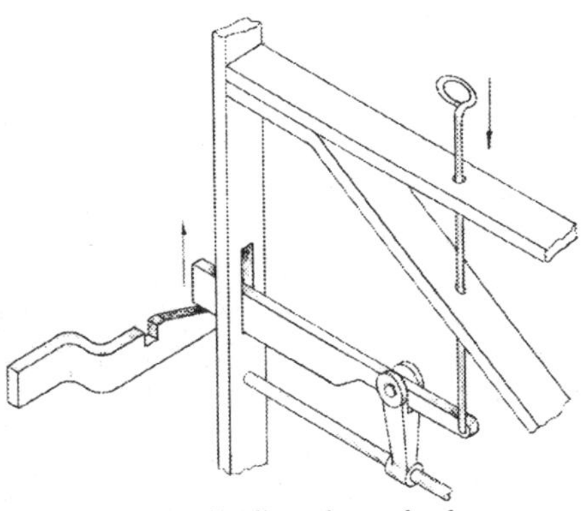

Bridle path gate latch.

Some gates are made to open in either direction and they have a sloping lead on both sides of the keeper plate to allow the latch to enter from either side. These gates always tend to close of their own accord owing to a double notched bar at the bottom of the gate. An unusual and more expensive self-closing device is based on using an edge cam. A roller is set into the under surface of the gate's end stile and this rides on the cam so the gate rises as it is opened and falls, by gravity, to close.

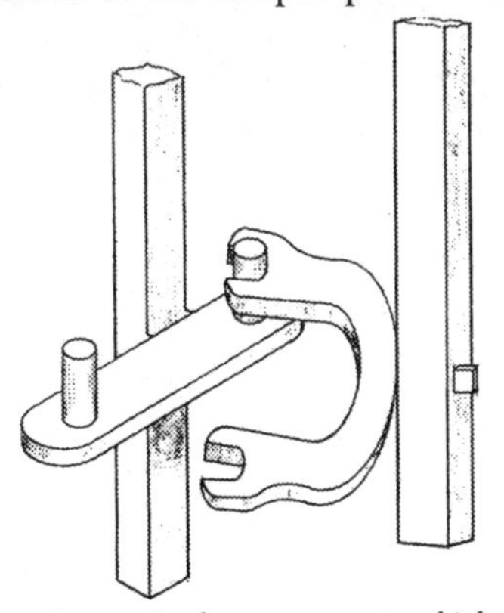

Automatic closure on a gate which opens in either direction.

Finally, the alternative to an arrangement involving a pivot is the sliding bar latch which can be a plain and simple design or enhanced by the blacksmith to add an aesthetic appeal.

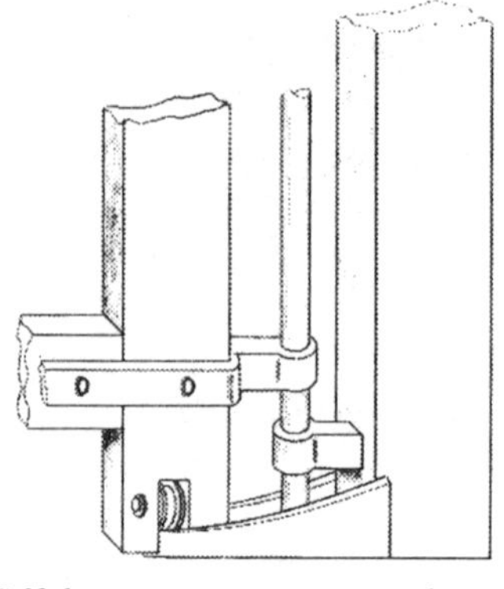

Self closing gate running on an edge cam.

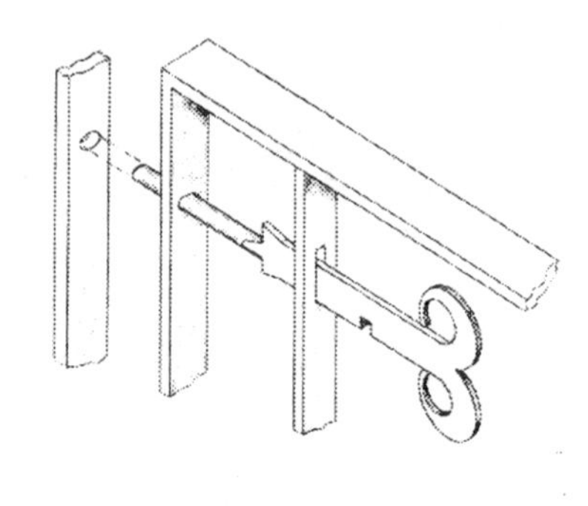

Sliding latch with a locking notch.

These comments on door and gate closure devices have been an attempt to show the uncommon nature of common objects. The overriding feature which appears is the degree of craftsmanship involved, and if one looks further than the catches on doors to the handles and knockers there is a wealth of styles from the crude to the beautiful. Such work is particularly fine on civil and ecclesiastical buildings on the continent of Europe.

Ornamental door furniture.

5
Standards and Measurement

From earliest times there has always been a need to measure and then be able to compare as a means of communication. It became evident in manufacturing that making identical items which needed to be interchangeable meant that they must actually be identical, and this was only achieved by measuring using a standard benchmark which everyone understood. A good example was in lock making, where a broken part could not be replaced 'from stock' until standardisation was established and identical parts were made. Similarly, roofing slates were of any shape and size until 1750, when a range of sizes were made standard and identified by being named after female aristocratic titles. The largest was Queen – 30 inches long; then Countess – 20 inches by 10 inches, etc. Obviously, replacements for broken tiles could easily be supplied and they would fit.

John Wilkinson, the ironmaster, had an interest in lead mining and this gave him the opportunity to make lead pipes. He made a pricing agreement with the Coalbrookdale Company and agreed upon a uniform range of pipe sizes. By doing this, Wilkinson was recognising the need for standardisation in products made by different manufacturers about half a century before such standardisation in products became self-evident.

Although methods of linear measurement have existed since antiquity, it was in 1305 that King Edward I defined our imperial system. He ordained that 'three grains of barley, dry and round, makes one inch, twelve inches make a foot'. This measure of barley corns persists to this day in measuring shoe sizes, with a base starting point of four inches and one barley corn (one third of an inch) for each successive size. Another example where natural objects were used as a basis for measurement is the carob nut as a gauge for precious stones, and this is the foundation of the carat unit used today as a measure for diamonds.

Many imperial measures are based on ordinary usage. A furlong (one eighth of a mile) represented the length of a ploughed furrow before the oxen team pulling the plough needed a rest. The rod, pole or perch (about five and a half yards) was the distance from one ridge to the adjacent one in the ridge and furrow strip method of cultivation and was the amount of work an oxen team could do in a day.

The rod, pole or perch measure was used in the building of churches and cathedrals where the master builder, usually itinerant, would carry his rod and use this in the construction, which meant that the proportions of the building were in units of five and a half yards. There are many examples, both in Britain and on the continent of Europe, where the width of the nave is exactly two rods, or eleven yards. One exception is Notre Dame in Paris where the nave is three rods wide.

Early devices used to make accurate measurements were simple. The Egyptian Libella is a good example of ensuring a horizontal surface and it was used for many centuries until superseded by the spirit level. The aqueducts built across Europe by the Romans needed to have a gradient sufficient to ensure a steady flow of water and this may be represented by the difference in height of a metre length of wood of the thickness of a playing card from one end to the other. Such a gradient was repeated over many miles across difficult terrain. An angle of 90 degrees is easily created using a triangular

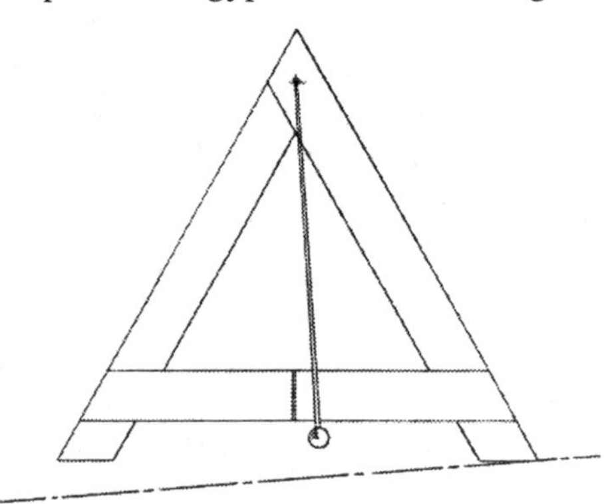

The libella – superseded by the spirit level.

The Pont Du Gard aqueduct in France.

template having the sides in the proportion of 3:4:5 (using the Pythagoras theorem). A simple way, used in building, to make a straight line is to stretch a taut, chalked string adjacent to the work and when 'snapped' it delivers an absolutely straight chalk line.

The relationship between the regular hexagon and the circle was understood in ancient times. In the trade of coopering (making barrels from staves of wood) the cooper would cut a groove around the inside of a barrel just below the top edge to receive the lid. In order to accurately obtain the exact diameter of the lid he would step around the bottom of the groove with compasses and by trial and error mark exactly six divisions. This dimension was the radius of the wooden lid, which was then hammered into the groove to make an airtight seal to the barrel.

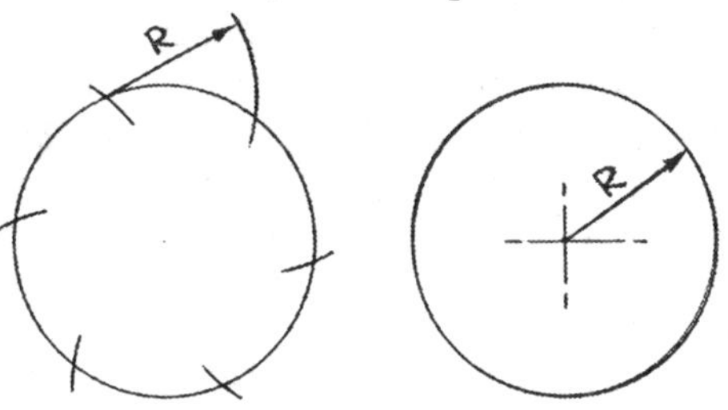

Method of determining the diameter of the lid of a barrel.

In the early years of the 19th century every trading centre in Europe had its own local foot and inch measurement and this varied from the English at 0.93 inch to the Russian at 1.75 inches. The English fathom of six feet was 1.8 metres whilst the Italian fathom was 0.70 metre and the French one somewhere in between. Trading between the nations for large commodities obviously demanded great care in negotiating prices.

The desire to cling to familiar and traditional methods is a human trait well demonstrated in the adoption of standards of measurement. The Paris garment trade was still using the French inch sizes well into the 20th century, some 80 years after the adoption of the metric system. Much earlier, in 1670, it was proposed that the metre be one minute of the circumference of the Earth and then, more than a century later, it was decided that the metre be one ten millionth part of a quarter of the circumference of the Meridian passing through Paris.

In England the standard yard was established in the 15th century and started life as a length of brass bar of octagonal section with three divisions representing feet and these then subdivided into 12 to signify inches. Eventually the yard standard was designated by the distance between two lines engraved on gold blocks inserted into a bronze rod. It is interesting to note that, whilst a microscope could locate the lines on the standard, it was not possible for engineers to accurately transfer the standard for use in manufacturing and they proposed using pins and faces. The lawyers and astronomers disagreed, but eventually the more pragmatic view prevailed.

In Trafalgar Square, London a number of length standards in the form of brass strips were set into the masonry of the north parapet. This provided an easy way for the populace to compare actual purchased lengths of cloth with that ordered from the nearby market. In Italy a similar arrangement existed and if on checking against the standard measure a customer found that they had been 'short changed' they could inform the authorities, who would then visit the offending market trader and destroy – *bancarotta* – the business. From this word comes our term bankrupt.

Also in Italy, there are reliefs of roofing tiles set into the walls of official buildings so that a person could check that his roof conformed to the accepted standard, and this avoided any arguments between neighbours.

At ports around England the government relied upon a duty paid on the import and export of all goods and services as an important source of income. Every conceivable item was subject to the duty, from perishable goods to the conveyance of coffins. The officer responsible for collecting the dues would calculate the amount and for this purpose he used a slide rule.

Slide rules were logarithmic calculators, very easy to use and surprisingly accurate. Making them in the 17th century required skill to ensure that the graduations were true and gave consistent readings. Sometimes the sliding scales were graduated on three or

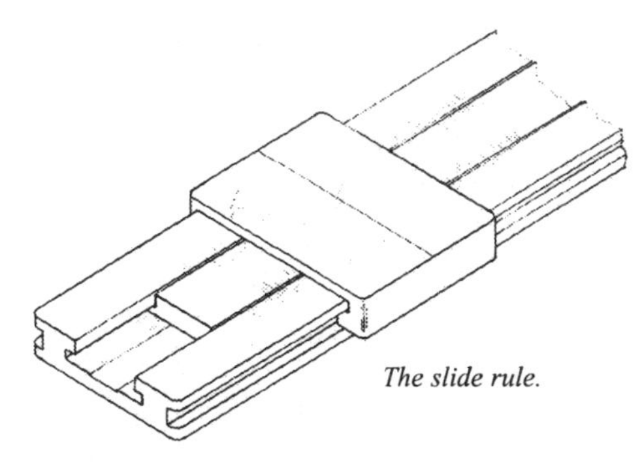
The slide rule.

four faces of a square section and engraved on the rule; separate from the graduations would be mathematical data and factors to help with the calculations. Slide rules were made dedicated for a particular trade.

When a carpenter wanted to make an octagonal section (eight sided) from a square (four sided) piece of timber he used a rule knows as the 'eight square lines' or mast-maker's scale. The scale had designations E and M (edge and middle) and produced a true octagon. The alternative was by trial and error or using trigonometry which took longer.

The Roman rule was a wooden stick divided into feet; palms; twelfths or unciae (inches) and digits (finger widths), but with the establishment of the inch and foot the demand for accurate rules really started in the early 19th century, when the manufacture of machine tools was gathering momentum and Joshua Routledge designed the engineer's rule. Initially this was intended to be used to calculate the weight and volumes of various materials and the duty of steam pumps, but presumably it also served as a linear measuring instrument.

Pattern makers could not use a conventional rule to measure their wooden patterns since an allowance for contraction had to be made to guarantee that the casting would be to the correct dimensions. The rules allowed for a contraction of 0.1 inch per foot for cast iron.

The making of rules initially involved engraving the graduations using a marking knife guided by a tri-square and copied from a master rule placed alongside. A good worker could mark in excess of 500 lines in ten minutes and presumably they were accurate. Once the graduations were marked on a boxwood blank a mixture of carbon black and oil was rubbed in and a coat of polish applied to finish it off.

Pierre Vernier published his scale in 1631 for the accurate reading of scientific instruments, quadrants, etc. His design was improved by William Gascoigne in 1638 and adopted for general engineering use when the technique for engraving the scale had evolved. It was only superseded in the late 20th century by electronic direct reading. The Vernier scale is most commonly attached to a calliper, an instrument having a pair of jaws, one of which slides along the scale and measures the distance across the jaws, which are positioned over the item being checked, to an accuracy of 0.001 inch (one thousandth of an inch). Vernier scales were also used on protractors, and angles could be measured to an accuracy of five minutes of arc (or one twelfth of a degree).

Principle of the Vernier scale.

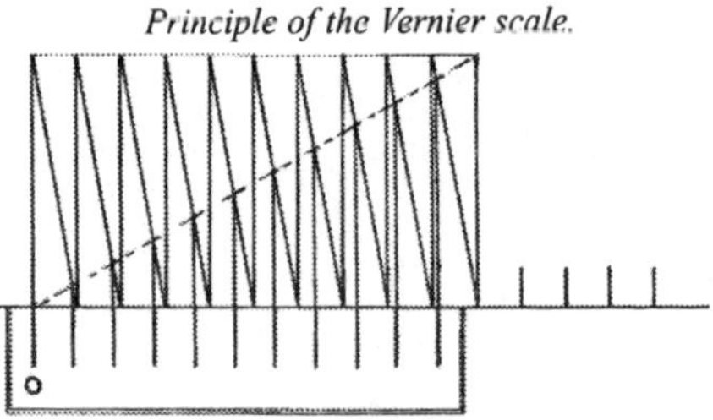

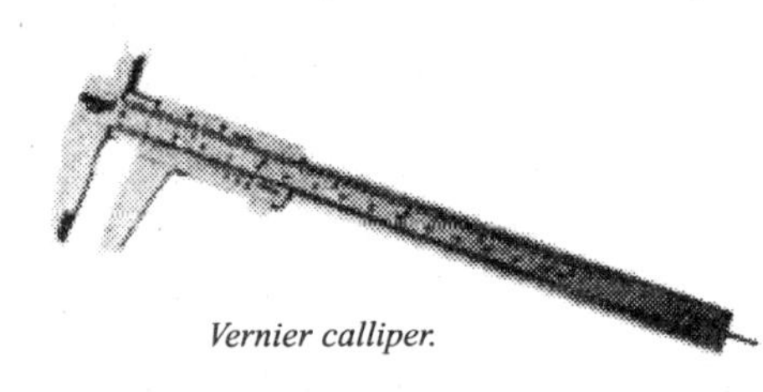

Vernier calliper.

The micrometer has consistently been the most commonly used measuring instrument and when fitted with the Vernier scale can record to an accuracy of 0.0001 inch (one ten thousandth of an inch). When engineers talk of measuring to a tenth, this is the accuracy that they mean.

Vernier protractor.

The earliest micrometer for engineering measurement was made in 1819 by James Watt, and shortly afterwards Henry Maudslay used his expertise in accurate screw cutting to produce the 'Lord Chancellor', a micrometer he considered to be the absolute of all verdicts in the ultimate Court of Appeal. The graduations on the hand wheel of the instrument indicated thousandths of an inch and it was capable of measuring up to eight inches with accuracy to within five thousandths of an inch.

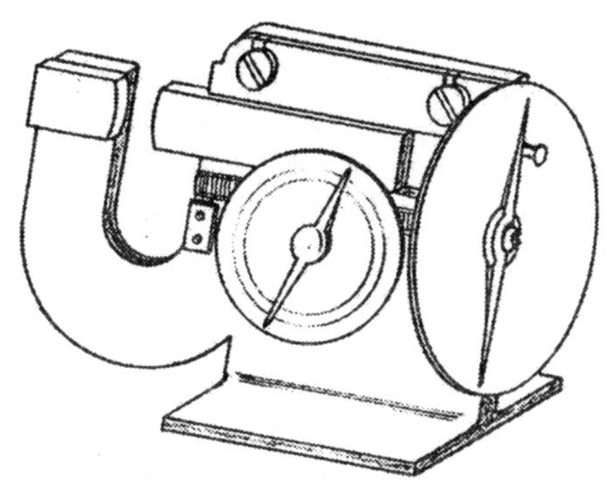

Watt's micrometer, 1819.

The traditional method of measuring was by callipers and rule until the micrometer, being a direct reading device, took over and made measuring easier. Greater skill was required using callipers to get the correct 'feel', the sensory pressure, to ensure that the calliper jaws opening were exactly the same as the item being checked. Callipers were comparators, a means of comparing an item against a measurement on a rule or to produce identical items.

Micrometers became invaluable as general measuring instruments but there were still applications where the calliper found favour. One example was the gunner's rule which was used to measure the calibre (diameter) of shot, and another was to calculate the number of teeth required in meshing gears for clock making, called the sector. In both cases graduated scales on the calliper arms made checking easy without recourse to another measuring instrument.

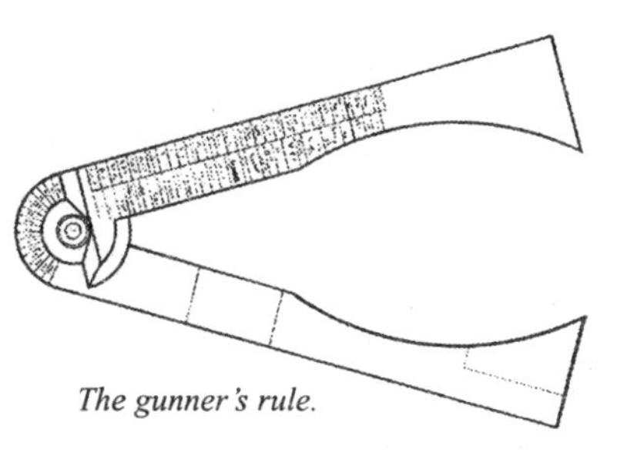

The gunner's rule.

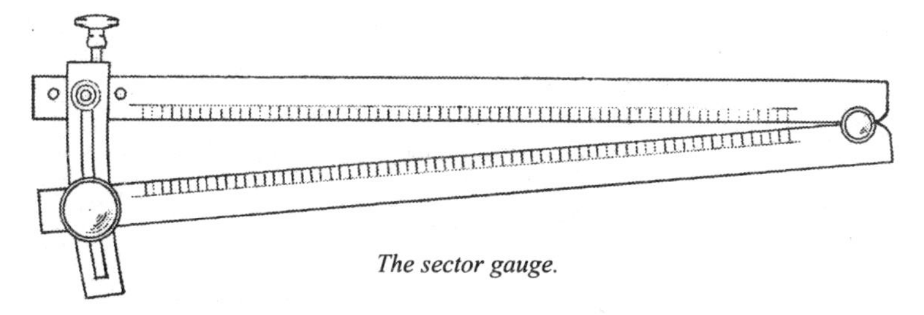

The sector gauge.

Continuing the theme where the natural world has influenced engineering applications, we turn to the problem of creating graticules – fine lines on glass – for observation in optical instruments. During the 17th century silk thread was used. Silk is a form of protein produced by the larvae of the moth Bombyx Mori and is a material well known when spun and used in textiles. However, the 'silk' produced by spiders to make their webs is finer and of more uniform thickness than the silk moth's product. It is also stronger and, as mentioned earlier, can allegedly stop a bee travelling at 20mph without breaking. Spider 'silk' was used in the eyepiece of a telescope for a ten line graticule when Greenwich Mean Time was determined and the world's prime meridian established at Greenwich. In more recent times a well known precision engineering company required a registration line on a transparent screen as an eyesight datum on a fine measuring machine. After having explored the possibility of engraving a very fine line they resorted to using the silken thread of a spider. The strand was fine, of uniform thickness, smooth edged, dark and strong. A variety of spiders could no doubt satisfy this specification but the very best specimens were found inhabiting the bicycle sheds of the company.

In the manufacture of machine tools it was vitally important to be able to create absolutely flat surfaces on metal. This was achieved by Joseph Whitworth, whilst working for Henry Maudslay, who created a set of three surface plates absolutely flat by checking and adjusting each one against the others. These plates then became the masters against which all work was checked.

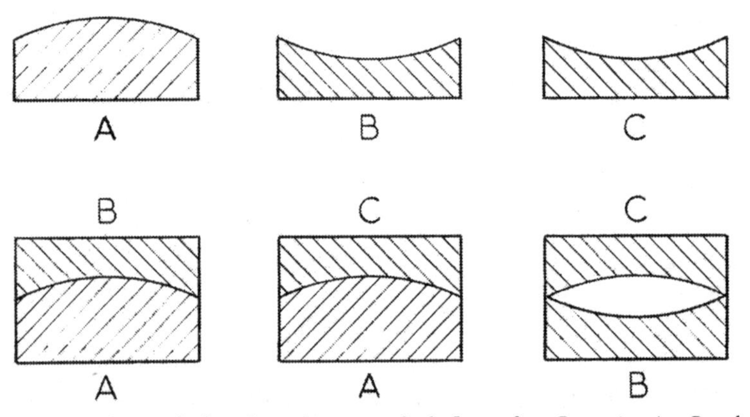

The three plate method used to achieve a perfectly flat surface. By mating A to B and then A to C, any errors in flatness are only revealed by mating B to C.

Measuring flatness developed and in the 20th century the accuracy of gauges was accomplished by placing an optically flat glass block on the surface and noting the number of fringes, each representing one wavelength of light (0.00001 inch). This technique is known as interferometry.

The master standard generally used to check measuring instruments was the set of slip gauges. This consisted of 81 blocks which, when used in combination, could be used to measure from 0.1001 inch to 4 inches in units of 0.0001 inch. Once again, credit for this innovation must go to Whitworth in 1842, although the Swede, Johansson, claimed ownership some 54 years later. Slip gauges became known as 'Jo' blocks and perhaps it would be more appropriate to call them 'Joe' blocks – after Joseph Whitworth – but then credit is not always accorded to those who most deserve it.

6
Materials

The craftsman has a whole range of material resources – animal, vegetable and mineral – with which to demonstrate his skill and creativity. In engineering the mineral resources predominate and iron is the most widely used.

Cast iron, when poured from the furnace, runs into gullies from a main channel and the ingots so formed are called pig iron as they resemble piglets feeding from a sow. Such iron is very brittle and used to be worked with a helve or tilt hammer to remove the slag. It was also necessary to reduce the carbon content left due to the charcoal fuel, and this was achieved in the 18th century when coke was used in smelting and provided the necessary increased temperature. Using coke also overcame the problem caused when coal was used and the sulphur united with the iron as an impurity. It was perhaps fortunate that coke, as an alternative fuel, was developed at this time because there was a shortage of timber, particularly oak, to make charcoal, due to the intensive activity in shipbuilding during the Napoleonic Wars.

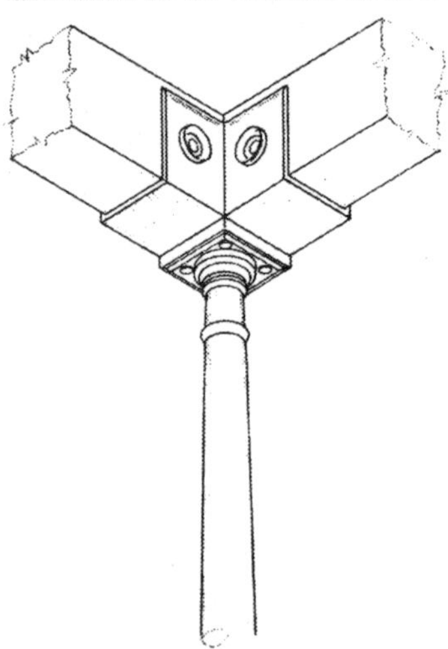

Cast iron column with a suggestion of classical influence to give aesthetic appeal.

Cast iron is strong in compression and will absorb both vibration and shock loads, which made it ideal for the bases in the developing machine tool industry. John Smeaton introduced the first cast iron shaft for watermills in 1769; it was capable of transmitting greater power than the timber one it replaced and was much smaller in diameter. In buildings there was generally little cost advantage in using cast iron for beams, but for textile mills it was an opportunity to improve the design by the lower weight/strength ratio.

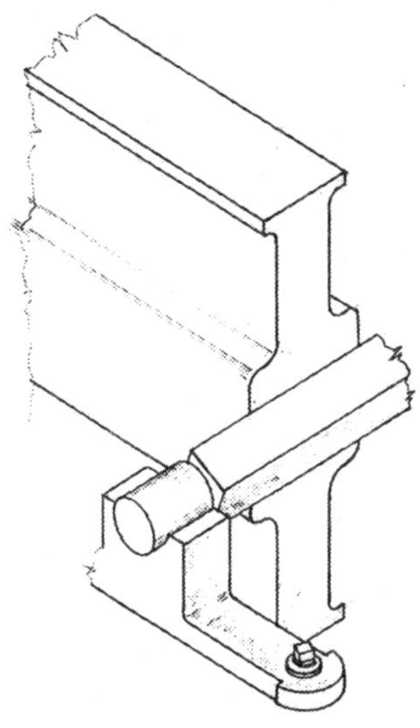

The weakest point of the beam in a beam engine, where the pivot is situated.

Castings could be made to look attractive; there was no limit to the degree of ornate decoration, and once the pattern maker had produced the wooden pattern for the mould, many identical castings could be made. Due to its brittle nature, cast iron is easily fractured and smooth changes in profile and a uniform cross section are necessary to reduce stress points. A regular section also helps to avoid porosity and 'blow holes', which can be caused by an incorrect temperature on pouring, generating internal pockets of air not normally visible from the outside. The cast iron beam of a beam engine would normally work well for a period up to 100 years but there have been occasions when one has broken resulting in injury and death.

Cast iron is little affected by the corrosive action of salt air. This is very evident in the coastal towns where piers, shelters, bandstands and railings at the seaside survive and are also testimony to the lavish decoration which adorns these objects to delight the aesthetic senses of the holidaymakers.

Cast iron railings.

When cast iron is worked using a helve or tilt hammer the result is wrought iron, where the carbon content is reduced to 0.1 per cent. Wrought iron has a slag impurity which resists corrosion and it is also stronger than cast iron. Beams of wrought iron have been used in structures when the cast iron beams collapsed. Good shock resistance and its excellent hammer welding properties are equal to modern welding techniques.

Steel has become the most widely used material of the ferrous (iron) types, especially with the addition of alloying elements, and finds universal application where its unique properties are needed. Steel could be heat treated to improve the properties, specifically by hardening and tempering to make springs. In the 18th century, suitably treated steel was used for the cutting area on edge tools. It was important to get the hardness of a cutting tool right; if too soft it would not hold the sharp edge and if too hard there was the possibility that it would shatter.

The first half of the 19th century was the age of wrought iron; the second half that of steel, the basic form being mild steel. Unfortunately corrosion resistance was (and is) poor, and a finger placed on a piece of mild steel will show as a rusty fingerprint within 24 hours.

The traditional metal used in bell founding is bronze, but in the 1850s there was a demand for a cheap alternative and steel was picked to make bells. The result was a pitted surface, corrosion susceptibility and deterioration in tone, so bronze consisting of 77 per cent copper and 23 per cent tin was restored. It was not easy to transport a bell weighing up to three and a half tons by horse and cart from the foundry to the church, so the founders became itinerant, moving around the country and digging the pits for casting on site. Bells were cast to be slightly sharp in pitch and metal was then removed to achieve the required tone. It was possible to tune a bell to within one frequency per second and this was considered good enough for the human ear!

The balance of copper and tin was changed to make different types of bronze for different applications. In the same way different types of brass were made by adjusting the proportion of copper to zinc to alter the properties and make the metal appropriate for the job it had to do.

It is recorded that the first alloying of copper and zinc to produce brass in England took place in the Wye Valley in 1565.

Previously, sheet brass was imported from the continent of Europe, and the evidence is present in the memorial brasses dedicated to the nobility of the 15th century in English parish churches. The most notable of these show exquisite engraving and minute detail in the designs. One example of the use of brass in the early days comes from paper-making, where the gauze was made of brass wire 0.008 inch diameter with 60 holes to the inch on a loom 100 inches wide.

Timber has been used extensively, since it is much easier to manipulate than metal. The ability of certain timber to withstand water saturation has been put to use in, for example, making clogs of alder wood. Elm was used where permanent immersion was inevitable, for instance in water pipes and the lock gates on canals. Generally wood will absorb water and, therefore, increase its mass. The massive timber beams on beam engines prior to1800 protruded from the engine house and were exposed to the elements. A story relates how this situation caused some concern at a mine in Cornwall when the engine was idle for a long period, and in dry weather the beam was in the up position, but in wet weather it became heavy and descended of its own accord. So, as the weather changed the beam oscillated up and down, no doubt to the consternation of the local populace.

Mahogany was used as the base (bed) for lathes before cast iron took over due to its stability.

The combination of materials such as ebony, rosewood and mahogany with brass, enhanced objects and gave an appeal more appropriate to the drawing room than the utilitarian use intended. Brass has invariably been combined with mahogany to make scientific instruments and imparts prestige, creating respect and esteem. Both rosewood and cedarwood give off a sweet scent and can be detected even after a long period of use.

The activity of the gunsmith in chasing and inlay work on gunstocks, using Mother of Pearl and ivory, was particularly fine. An unusual use of animal material was to control the swing of the balance wheel in clock escapements, where hog's bristles limited the rotation and dampened the energy. Unfortunately, the bristles varied in their ability to do the job and also deteriorated; the advent of the alternative metal replacement ultimately solved the problem.

One interesting material was papier mâché, which was used to make furniture; often highly ornate and heavily lacquered. The technique was perfected in the far east and arrived in Birmingham where the trade flourished, but sadly has now been lost completely.

A skill which is akin to sculpture was the dressing of mill stones. Various configurations of groove pattern were evolved over the years, all with the objective of achieving effective grinding of corn. Millstones which rotated clockwise were easier to dress using a bill (chisel), and therefore the sails on the windmill would rotate anti-clockwise. The French bur stone was favourite but the millstone grit from Derbyshire was a little nearer home, and it is interesting to note that even today the hillside near Stanage Edge is littered with discarded millstones. Possibly the result of over production when the industry declined.

When considering the relative strength of different materials it is interesting to compare the length of identical sizes at which a fibre will break. The longest is diamond crystal, followed closely by carbon fibre, with spider's web giving a good result; the poorest outcome is shown by steel.

Finally, it is remarkable to reflect on what was achieved with all sorts of different materials during the 18th and 19th centuries; how the craftsmen used their unique properties to the full and had the skill to overcome the reluctance of materials to be manipulated, maintaining respect for the nature of the substance and always working with it rather than fighting against it.

7
Motive Force

7:1 Transport

Transportation during the 18th century was by four methods – a road network, some canals, coastal transport and tramways. All these systems were used with varying success but we now know that the most enduring (at the time) was the railway. There were a number of innovative suggestions, none of which was found acceptable or possible. H.R. Palmer proposed a monorail supported on pylons which would overcome the problems presented with gradients on the ground with a railway. George Cayley designed a system which we now know as caterpillar tracks, where the vehicle created its own road surface.

At the end of the 18th century efforts were made to adapt the stationary steam engine to enable it to pull wagons on a track or drive a coach on the roads instead of using horses. The problem was to build a vehicle which was small and manageable and yet capable of carrying the power unit, an external combustion engine with all its associated equipment.

A number of attempts failed for various reasons and it was in 1801 that Richard Trevithick built the first steam carriage which was able to run on roads. To mark the bicentenary of this achievement a working replica was built for the Trevithick Society. This has given us an insight into some aspects of the original design. Steering was by a tiller attached to the axle tree holding a pair of small wheels, no allowance was included for turning and the resulting erratic course was not helped by a minimum speed of eight miles per hour. When stationary the

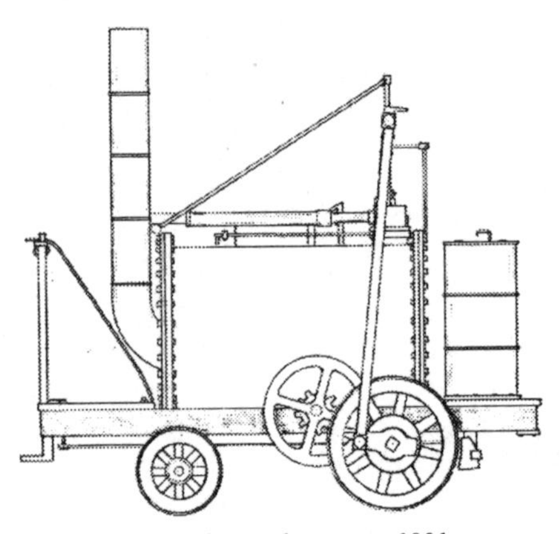

Trevithick's road carriage 1801.
Outline of 2001 replica.

connecting rod to the drive wheels usually stopped at bottom dead centre and in order to get it started the passengers had to jump off and manually rotate the wheels so that the engine could drive. One interesting aspect is the method of checking the water level in the boiler. Two taps were positioned on the boiler, one at the high water level and the other at the minimum level and by opening the upper tap when filling, it was possible to get the water level just right. A more obscure method was to suspend a stone on a measured rope from the filler and to sense the reduction in weight of the stone as it became immersed in the water.

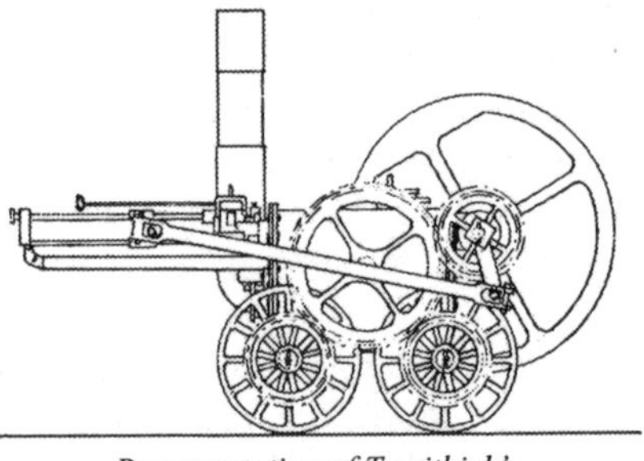

Representation of Trevithick's Pen-y-darren locomotive 1804.

Trevithick continued, despite setbacks, and built a carriage for the London streets and a locomotive for the Pen-y-darren iron works in South Wales. All the innovations were not developed, owing to a lack of attention to the correction of small problems and some deficiency in business acumen. To be fair, the failure of the Pen-y-darren engine was due to the breakage of the cast iron tramway under the five ton weight of the locomotive, and since this was a competition Trevithick was deprived of the prize.

Driving gear on Pen-y-darren locomotive.

Another early success in building a steam road vehicle was in 1784 by William Murdock, who was employed by James Watt as his engine erector in Cornwall. Murdock spent his leisure hours constructing a working model of a locomotive. He tried it out, successfully, and managed to frighten the local pastor in Redruth. James Watt, on hearing of the success of the steam carriage, apparently persuaded Murdock, through his business partner Matthew Boulton, to discontinue further development on this project as it would detract from his work for the company.

Murdoch's road carriage, 1784.
Sketch of replica.

James Nasmyth, when he was 19 years old in 1827, was asked by the Scottish Society of Arts to construct a steam carriage to carry eight passengers. This was successful and carried out runs of four or five miles. No commercial value was attached to this vehicle, possibly due to the atrocious condition of the roads at the time. The Turnpike Acts brought only a gradual improvement and there was the view that the horse had served well and could continue to do so. There was also the imminent prospect of travel by rail, and it is interesting to speculate how our lives would have changed had it been possible to improve the roads at that time rather than developing rail travel.

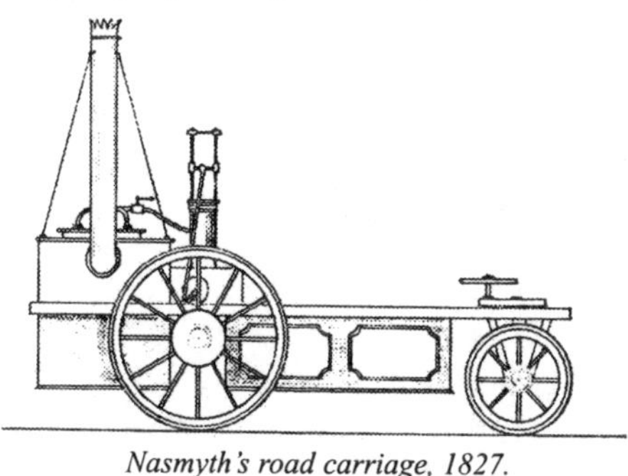

Nasmyth's road carriage, 1827.

Rails were initially made of wood resting on wooden sleepers, then, to prevent wear on the timber, strips of iron were attached to the rolling surface. Development continued with rails of angular section to keep the wheels in place, resting on a stone sleeper. The next step was to make the 'double mushroom' rail for flanged wheels and this continues to this day. Benjamin Outram was responsible for aspects of rail design and his 'plate ways' used in collieries were known as 'Outram Roads', in turn this was shortened to 'tramways'.

The sleepers were positioned under the joints in the rails and the unsupported centre of the rail was made 'fish bellied' to give strength. Such rails in cast iron were adequate for horse drawn freight but with the advent of the heavy steam locomotive, the weakness of cast iron was overcome by using wrought iron, and this was patented in 1820 by John Birkinshaw who produced rails in lengths of 18 feet.

As steam locomotives were being developed it was thought that the smooth wheels on the smooth rail would turn without 'biting'. John Blenkinsop took out a patent in 1811 for a toothed wheel on the locomotive to engage with a rack on the rails to provide traction. The engineers disagreed on the benefit of this system and although it was tried, at great expense to make, it failed due to stones becoming trapped and the gear teeth breaking. George Stephenson was able to prove that smooth wheels, with the help of the weight of the locomotive, had sufficient adhesion to drive the engine. Wheels on locomotives became six to eight feet in diameter to enable a greater distance to be covered for a given power output. However, adhesion is reduced by using such large wheels, so a compromise was achieved by having smaller wheels and more of them.

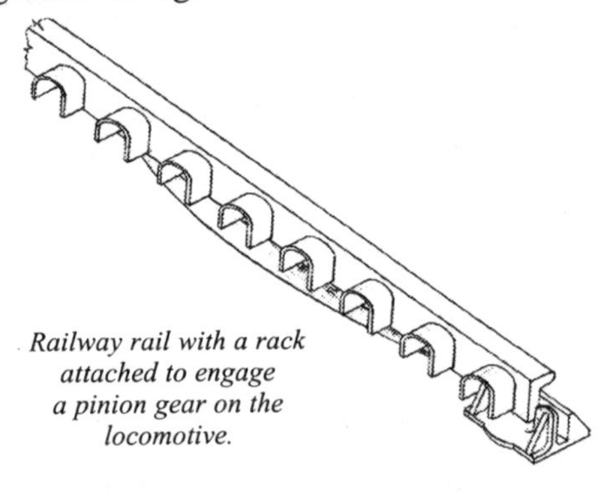
Railway rail with a rack attached to engage a pinion gear on the locomotive.

A gauge of rail between three feet nine inches and five feet was thought to be optimum for a horse economically hauling a load. The wheel base was always longer than the axle length, so that no more than one quarter of the wagon load was bearing at any one time on a single rail. In coal mines the gauge was two feet three inches and the early locomotives frequently broke down, so as a precaution the horses which previously hauled the wagons followed behind in readiness to regain their former duty if required. Richard Roberts in Manchester made locomotives with a gauge of six feet two inches and considered this to be the optimum based on practical considerations.

He argued that the power of a locomotive is limited by the size of the cylinder producing the steam and this in turn was determined by the spacing of the side frames. These were constrained by the gauge (the space between the wheels). The gauge he proposed would enable adequate power to be provided, make it easy to fit the cylinder and facilitate repair and replacement.

Eventually the railways were made in two gauges. Stephenson used four feet eight and a half inches in the north and Brunel made his to seven feet and one quarter inch in the south of England. After a protracted dispute Parliament decided that the narrow gauge would be the standard, and Brunel renewed his permanent way in 1892 over one weekend using a team of 4,700 'navvies'. The term 'navvy' derives from the workers who built the canal system – the navigation – some one hundred years earlier. Before the standardisation of the gauge, travellers between north and south had to change trains en route in places such as Gloucester. Brunel had convincing arguments in favour of the broad gauge, especially for safety reasons, but it was not to be and the narrow gauge, which had in part a historical connection based on the Roman chariot, triumphed.

The success of the railways and the failure of the road carriage was primarily due to the relative smoothness of the surface over which the vehicle ran. Rails were hard and smooth, no steering was required, a long train of carriages could be hauled. On the other hand the steam carriage was hampered by the atrocious state of the roads and

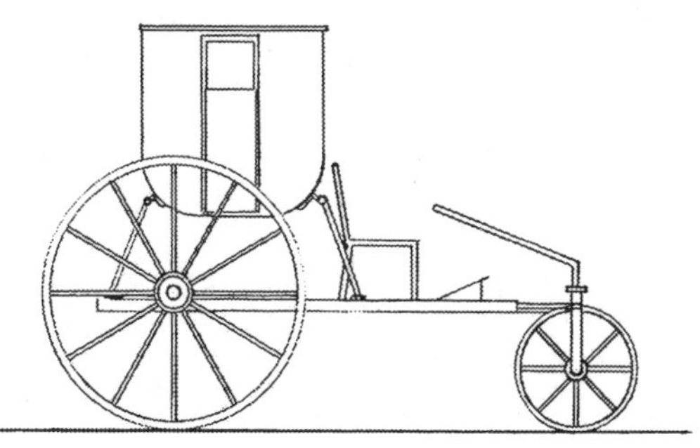

A steam engine replaces horses for power but the stagecoach style remains.

f

constraints of weight and space. Road transport continued with the stage coach, undertaking long journeys around the country drawn by horses. Eventually steam power superseded the horses and the design of the new vehicle bore a strong resemblance to the previous stage coach.

One novel form of rail transport put into use by Brunel in Devon was the atmospheric railway. It worked by having a tube placed on the ground between the rails which was connected to a stationary steam engine at the end of the track. A vacuum was created in the tube and a piston in the tube was connected to the train. An open slot in the tube allowed the connection to the train and this was covered by leather flaps which were usually displaced as the train passed. Unfortunately, the leather of the flaps became stiff at low temperatures and lost the ability to seal effectively. Further, rats were partial to the leather and made a meal of it; consequently the project was abandoned. Previously the atmospheric railway had been tried in Dublin; and William Murdock had experimented with compressed air and conceived the idea of transmitting letters and parcels along a tube exhausted by an air pump. This was later practised with success by the London Pneumatic Dispatch Company. Perhaps it should be noted that the principle of the slotted tube with piston protruding was applied to pneumatic cylinders used for control systems in engineering in the late 20th century, however, the seal medium was now a sophisticated plastics material brush, which was not available in Brunel's day.

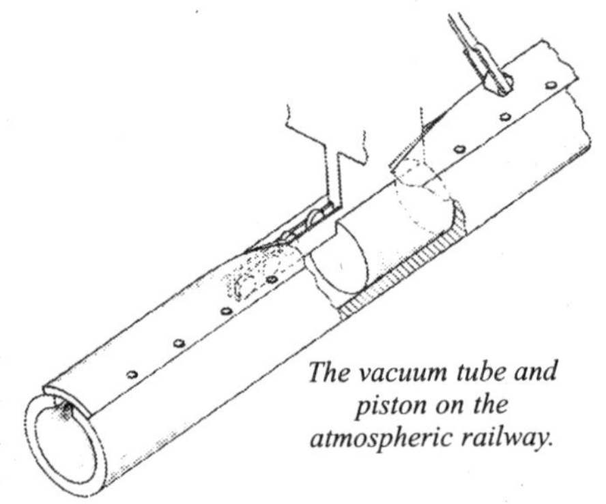

The vacuum tube and piston on the atmospheric railway.

Until the railways arrived the canals provided a perfectly adequate service for carrying goods around the country at a sedate pace. The transportation of coal was the initial reason for the canals being built, but other heavy mineral materials were carried to support the growth of industry. The site at Coalbrookdale in Shropshire, now acknowledged as the birthplace of the Industrial Revolution, was ideally placed by the River Severn, and other centres developed as canals were constructed.

In the Manchester area the transport of coal was by a combination of pack horses carrying baskets and boats hauled by men (horses and mules were not used until the canals were built). James Brindley was responsible for constructing the first aqueduct in England over the River Irwell at Manchester. He was ridiculed, when the proposal was made, for suggesting that boats could travel through the air! This was the penalty he had to pay for originality and genius. None of his detractors was aware, apparently, of the existence of aqueducts on the continent of Europe or for that matter the relationship between the weight of a boat and the weight of the volume of water displaced. Brindley also created the canal lock to enable boats to overcome a gradient.

The means of making a canal watertight was by 'puddling' and this was done by using a mixture of sand and clay and manipulating it, like kneading dough in breadmaking. The resultant concoction was then applied to the bottom and sides of the prepared canal in layers building up to three feet thick. Brindley apparently gave a practical demonstration of puddling before a committee of the House of Commons in the 1760s to gain approval for his canal construction. All the ingenious techniques which Brindley employed in his work were by his own original thinking, with no previous experience and no advice from foreign engineers in similar circumstances. We know today how well his work has successfully stood the test of time. There is a danger that the work of some very competent engineers is overshadowed by their more illustrious associates. Such was the case where Brindley's assistants went on to build canals in their own right, and credit should be given to Hugh Henshall [1734-1816] and Josiah Clowes [1735-1794].

Birmingham gained prominence when the connecting canal to the River Severn was begun in 1769. In the early 1800s it took longer for a bale of cotton to get from Liverpool docks to the Manchester textile mills than it did to cross the Atlantic from America. Both the railways and the canals solved this situation.

The proposal to build the Grand Trunk Canal in 1758, to join the rivers Trent and Mersey to ease the problem of transporting goods and materials on the appalling roads, was met with opposition. It was claimed that innkeepers would become bankrupt; packhorses and their drivers deprived of employment; towns bypassed thereby losing trade; landowners would be obliged to forfeit their land; seamen put out of work and the coaster work superseded. In the scheme of things there has always been proposal balanced by opposition and we can only hope that when the differences are sorted out the resultant course of action is the very best compromise. In the case of the Grand Trunk Canal we now know, in retrospect, that trade benefited greatly and no doubt the opposition to the scheme went on to find new, and better, forms of employment.

The canal trade declined as the railways grew, and railway companies needed to have a vested interest in canals to ensure that the transport of goods by rail was successful. They erected signs where roads crossed canals which indicated the maximum load which could be safely supported without collapse. Interpretation of the various data on these signs was somewhat difficult and the driver of a vehicle about to use the bridge would no doubt either trust to luck in crossing or give up. This is assuming that he could read and understand the instructions and was thoroughly familiar with the intimate details of his vehicle's weight.

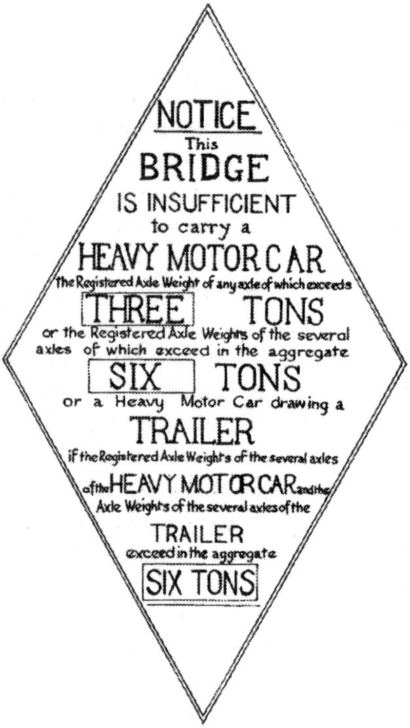

A road sign gives warning of the capacity of a canal bridge.

No canals existed in Cornwall, where significant manufacture of massive equipment for beam engines was taking place. Transportation from site to coast for onward forwarding by ship was done by teams of horses on atrocious roads across rough terrain. No coal existed in Cornwall and ships would take copper ore to the smelters in South Wales, owned by the Cornish, and on the return trip bring coal to Cornwall to fire the boilers of steam engines.

Credit for the first iron boat must go to John Wilkinson who, in 1787 floated this in the River Severn. Many doubters were disappointed when it floated! A number of iron vessels followed, the most famous being the SS *Great Britain* built by Brunel in 1843. This was the first ocean going vessel to be screw driven and constructed entirely of metal. Previously progress had been made in the propulsion of vessels using screw type propellers by John Stevens, Francis Petit Smith, John Ericsson and William Symington.

The early experiments in screw propellers, paddles and power by steam engines laid the foundations for marine engineering.

7:2 Power

The advent of the steam engine as a prime mover to replace a variety of traditional ways of working had an impact on industry and social conditions not experienced previously, in other words the Industrial Revolution.

Such changes reduced the drudgery of work and improved the quality of life. When a young man, James Nasmyth, made a four inch steam engine which was used to drive the lathes in a company in Edinburgh, the contemporary comment was that the busy hum of the wheels and the smooth, rhythmic sound of the engine, through some sympathetic agency had quickened the strokes of every hammer, chisel and file in the workman's hands so that the output nearly doubled.

Watermills were often positioned along the line of a river system and those most upstream would become starved of a water supply and, therefore, steam engines were used to pump water back uphill. Eventually it was realised that the steam engine could replace the watermill on site.

The steam engine does not tire, provided that maintenance is carried out, and can provide the steady concentrated power required over a long period of time. In Warwickshire some 500 horses were required to hoist water bucket by bucket, to drain coal mines, at a cost of £900 per year to feed them. The steam engine installed to replace them consumed £3,000 worth of coal per year, but this commodity was freely available on site! This was the Newcomen atmospheric engine, which was not as efficient as the subsequent engines developed by James Watt which were actually powered by steam (as against atmospheric pressure) with a separate condenser and required less fuel to run.

At the Soho Works in Birmingham of Boulton and Watt the main steam engine used the exhaust air to power small vacuum engines attached to individual machines, and this overcame the need for overhead line shafts with long flapping drive belts. Driving belts continued to be used in many places up to the mid 20th century, with overhead line shafting now driven by electrical power and the responsibility of making and repairing the leather belts falling to the 'beltman', fully employed in the art of cutting the leather, making the interlocking comb wire loops and fitting the cat gut securing pins. The belts were often quite long and when flapping as they powered, the danger to a machine operator of injury if they should break, or come off the pulleys, was potentially frightening.

James Watt, writing in 1782, defined the standard for power. He noticed that on a mill whim used to grind and rasp logwood, twelve horses were needed to drive the calendar. Each horse walked in a 24 feet diameter circle and made two and a half turns every minute, which equated to 60 yards per minute at a rate of 180 pounds per horse. Thus the amount of work done was 60x3ft x180lbs, i.e. 32,400ft.lbs every minute. He then applied this to a steam engine making 20 strokes of 6 feet per minute. So 32,400 divided by 120 gave 270 pounds for each horse, and since 12 horses were used the total load on the engine would be 3,240 pounds. An engine operating at 5 pounds per inch pressure of steam, with a piston stroke of 6 feet, at a rate of 20 strokes per minute, required a cylinder diameter of 29 inches.

Thus the standard for work was established using empirical data where one horsepower raises 32,400 pounds by one foot every minute. This was later changed to 33,000ft.lbs and remains the standard horsepower rating to this day.

8
Organisation

8:1 Learning the Craft

From monastic beginnings, to the Trade Guilds of the 19th century practising their crafts, a high standard was established which attracted commissions from royalty, statesmen and the Church. The starting point was the apprenticeship, where the skills were learnt and the knowledge base created to lay the foundations for someone to practise as a master craftsman.

The terms of the Deed of Apprenticeship in the 19th century were harsh relative to the more recent forms of formal training. The Deed was a legally binding contract between three parties: the employer, the apprentice and the father of the apprentice.

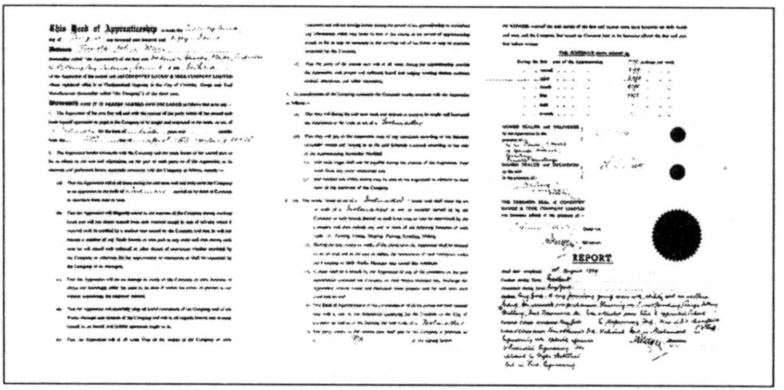
This Deed of Apprenticeship

REPORT

Example of a deed of apprenticeship. Note the three seals against the signatories.

The father would often pay a premium to the employer. Included in the Deed were the conditions under which the apprentice would work and apply himself for the whole period, normally seven years. Such conditions were:

- A promise to learn the art and craft.
- To keep the Company secrets.
- To do no damage.
- Not to contract matrimony.
- Not to play dice or cards or any unlawful games.
- Not to haunt taverns or playhouses.
- Not to absent himself.

The Deed, known as an indenture, arose from the practice of tearing the document in half, the father and the employer each having a record of the agreement and the 'indented' tear would match if the validity of the document needed to be proved at a later date in case of dispute.

James Nasmyth was anxious to work for Henry Maudslay but his father could not afford the premium required to place him as an apprentice. However, Maudslay was so impressed by Nasmyth's previous work that he took him on as an assistant. Some parents, led by false evidence of constructive skill in their offspring, apprenticed them for vast sums of money to an engineering firm. Maudslay stopped taking on such people as they were poor attendees, a bad example to competent apprentices and came wearing kid gloves! Young aspirants for engineering success would wear kid gloves since that was considered the genteel thing to do in 1820. Matthew Boulton declined to accept gentlemen apprentices despite large sums of money being offered as a premium. He preferred to set on the humbler sort of boy and orphans who could be trained up as skilled craftsmen, having first identified potential.

James Watt returned to his home town of Glasgow after failing to find work in London. He was then prevented from practising on his own account by the authorities as he had not served an apprenticeship within the borough, nor was he the son of a member of the privileged class. Watt found sanctuary in Glasgow University, beyond the influence of the establishment, where he went on to make precision instruments.

On completion of the apprenticeship, the now qualified craftsman could either work for another master as a journeyman or become a Freeman working on his own account. The day of completion was cause for celebration, the apprentice would be leaving the company and the ceremony of 'banging out' was performed. This practice continued in some companies until the mid 20th century but it was then purely a tradition as a job with the company was available if required.

Having accepted a young man with potential to be bound apprentice, an employer, himself a highly skilled craftsman, would encourage a desire in the apprentice to achieve perfection in his work. This was sometimes frowned upon by other workmen who saw quality as a means of depriving them of future repair or replacement opportunities. Another grievance was that if the young man became proficient he might take the existing workers jobs and, in their eyes, deprive them of their livelihood.

For a young apprentice to overcome the torment and ridicule to which he was subjected, he developed a strength of character which then gave him an advantage in later life. It has always been true that young people, and apprentices in particular, are the butt of practical jokes from their older workmates.

Conflict sometimes arose when journeymen, who had served a seven year apprenticeship, thought that they should be promoted over any other person. The employer, on the other hand, would look for natural ability and the personal qualities of zeal, cheerful temper and enthusiasm were more important to him than an in-depth knowledge of workshop practice. There was a view that someone who could acquire the necessary qualifications in two years was better than the person who was so stupid as to require seven years' teaching.

In the 1830s the stamp duty which attended the swearing in of a Freeman was one pound three shillings and sixpence, and it was considered extortionate after a period of seven years to have to find such a sum in order to practise a trade independently. As a Freeman, the craftsman was less vulnerable to trade recession than the journeyman working as a 'hand' who could easily become poverty stricken should work not be available at times of recession.

The concept of 'The Freeman' goes back many hundreds of years when nomadic people began to settle and to ply their trade. Until 1835 the Freemen were the civic leaders. They organised the trade guilds, maintained the quality of merchandise and controlled prices. Most Freemen obtained their position by patrimony but in certain cities, notably Coventry, it was only achieved by servitude, i.e. a recognised apprenticeship, satisfactorily completed.

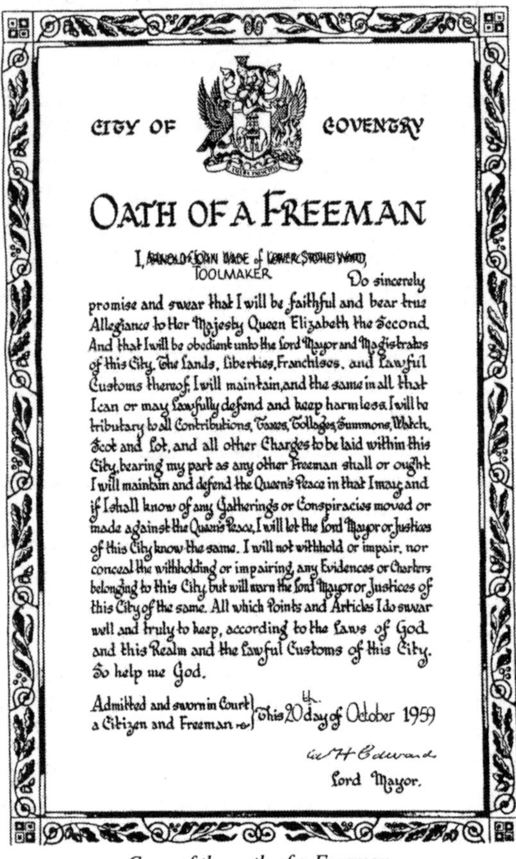
CITY OF COVENTRY

OATH OF A FREEMAN

I, ARNOLD JOHN WADE of LOWER STOKE WARD,
TOOLMAKER
Do sincerely promise and swear that I will be faithful and bear true Allegiance to Her Majesty Queen Elizabeth the Second. And that I will be obedient unto the Lord Mayor and Magistrates of this City. The Lands, Liberties, Franchises, and Lawful Customs thereof, I will maintain, and the same in all that I can or may Lawfully defend and keep harmless. I will be tributary to all Contributions, Taxes, Tollages, Summons, Watch, Scot and Lot, and all other Charges to be laid within this City, bearing my part as any other Freeman shall or ought. I will maintain and defend the Queen's Peace in that I may, and if I shall know of any Gatherings or Conspiracies moved or made against the Queen's Peace, I will let the Lord Mayor or Justices of this City know the same. I will not withhold or impair, nor conceal the withholding or impairing, any Evidences or Charters belonging to this City, but will warn the Lord Mayor or Justices of this City of the same. All which Points and Articles I do swear well and truly to keep, according to the Laws of God and this Realm and the Lawful Customs of this City. So help me God.

Admitted and sworn in Court a Citizen and Freeman } This 20th day of October 1959

W H Edwards
Lord Mayor.

Copy of the oath of a Freeman.

The Freemen grouped together in their respective trades to form the Merchant Guilds. This system, as originally constituted, failed due to restrictive practices despite the ethos of care and protection for the members. The Trade Union Movement grew and eventually exercised as much influence as the Guilds once enjoyed.

Sheffield in the 1760s was fettered by the guild system but in Birmingham any man could engage in business and the town developed for better – by those properly trained; and for worse –

by men who saw an opportunity to be successful by producing inferior products.

Sadly Birmingham became known as 'Brummagem' when used in a derogatory way to indicate the poor quality of the minority but, despite this, much excellent work was done with a phenomenal rate of production, giving the town the epithet 'The Workshop of the World'.

The trade guilds were closely linked to the Church of England so dissenters found it difficult to practise in places where the guilds operated. Regardless of this there were a significant number of masters and men alike engaged in the iron trade who were Quakers.

The Livery Companies in London are the fraternities drawn from the guilds and today there are 84 companies. The Livery Companies, like the Guild of Freemen, continue as social institutions more concerned with ceremony than having an overt effect on commerce.

The Compagnonnage is the association of craftsmen in France, started in the 15th century to improve the status of their trade. The training consisted of a long apprenticeship followed by several years travelling through France demonstrating and developing skills. No wages were paid and some ten years later a masterpiece in their chosen material would be produced to demonstrate fine craftsmanship, and they would then offer their services for reward.

In Italy, during the 13th century, the guilds (Arti) flourished as professional associations of merchants and craftsmen controlling trade, commerce and politics; with involvement in banking to generate a buoyant economy.

Craftsmanship is a thorough understanding of the material which is worked, an intuitive sympathy which can sense precisely what a material will and will not do. This understanding gives the craftsman mastery because it enables him to work with the material instead of against it. The question arises as to whether a craftsman is born with an innate ability or is it an acquired attribute? When working in wood against the grain the craftsman will sense a split in the wood before it occurs. A turner working at his lathe knows exactly how much material to remove without recourse to measurement. A fitter making an adjustment using a hammer knows exactly how much force is required to deliver a ‘two thou tap’, meaning to make a movement of two thousandths of an inch. An engineman knows by listening what is wrong with an engine when it is not running sweetly. Familiarity with a craft comes after many years of practice, to the extent that a sixth sense is developed through intuition which gives an awareness quite unperceived by the uninitiated.

Skill is the evidence of acquired ability.

8:2 Social Considerations

St. Paul wrote: *‘If any would not work, neither should he eat,’* and it is difficult in our present welfare state to imagine how true this statement was during the conditions which prevailed in the late 17th and early 18th centuries.

The craftsman had his skills to apply and could, therefore, be occupied, provided that work was available, but there was always the threat of the workhouse or poor relief if he should fall on hard times. If bankruptcy did occur his goods and chattels were sold, the exception being that he was allowed to keep his bed and the tools of his trade, so he would have the means of recovering his livelihood.

The position of the craftsman in the social order remained unchanged for centuries, since the reference in chapter 38 of Ecclesiasticus contained in the Apocrypha of the Bible:

> ‘The wisdom of the learned man cometh by opportunity of leisure; he that hath little business shall become wise; how can he get wisdom that holdeth the plough...and whose talk is of bullocks; The carpenter worketh night and day; The smith sitting by his anvil...the vapour of the fire wasteth his flesh; So doth the potter. All these trust in their hands. Without these cannot a city be inhabited. They shall not be sought for in public counsel nor sit on the judge’s seat. But they will maintain the state of the world and all their desire is in the work of their craft.’

Conditions in the Black Country – Dudley, Bromsgrove, Cradley Heath and Wolverhampton – during the early 19th century were particularly hazardous to the health of the people, where sulphurous fumes, smoke and devastated land was rife. The production of goods, however, was extraordinary, with a consequent generation of wealth but not necessarily for the humble craftsman.

The situation for the nail makers at Bromsgrove was appalling. Starvation was endemic. On the one hand was the prosperous owner and on the other the poverty stricken craftsman. Dishonesty and corruption were rife and the nailer was resigned to regret that his work was not valued but to live in hope of a reward in Heaven. Many of the craftsmen who ran a business on their own account were dissenters and this outward expression of non-conformity possibly reflects the mind of the inventor, and the artisan anxious to extend the barriers in the name of progress. The working man was ill disposed toward the Church of England with its liturgy, ritual and sermons delivered by a vicar who enjoyed a salary in excess of £1,000 when the ordinary man received about £25 per annum. An individualist religion, free of ecclesiastical ceremony, encouraged free thinking and self reliance. There was a history of strife prior to the 19th century where people suffered the Acts of Uniformity and were punished if they did not conform. Anyone not of the Anglican faith was prohibited from holding a position of authority in educational establishments or office in government.

Chain makers developed chain maker's cataract after long exposure to the heat from a forge. Lock makers in Willenhall developed a limp due to constant bending over a vice using a file; known as 'hind leg' as their body resembled a letter K.

As a result of the growth of the textile industry the population of Manchester increased by thousands and people lived in indescribable squalor. Despite the conditions there was a reduced incidence of disease due to the vast quantities of tea consumed, where water had to be boiled and the tea tended to inhibit disease.

Opportunities for activities during the limited leisure time were often spent in communal pursuits such as the formation of brass bands and choirs, particularly male voice, and today pockets of these pastimes continue in northern England, the Midlands, Cornwall and Wales. Areas where industry of an arduous nature flourished. Spiritual needs were met by a revival of the non-conformist denominations.

The need for education was felt, but there were reasons for not educating the very poorest working classes since they were needed to spend their entire lives working at a narrowly repetitious task, especially in the textile industry. Perhaps it would have been prudent to grant them literacy together with an understanding of moral values to live in an ordered society. However, there was a danger that if such people were educated they would then be in a position to challenge the views and attitudes of their employers. There is a parallel with this reasoning in the Apartheid system in South Africa which occurred during the 20th century.

Any organisation needs discipline to operate effectively and the following rules for office staff posted in 1854 outline the expected standard of behaviour, the punishment for transgression is open to conjecture:

1. Godliness, cleanliness and punctuality are the necessities of a good business.
2. The staff will not disport themselves in raiment of bright colour.
3. A stove is provided and each person to contribute 4 pounds of coal each day.
4. The calls of nature are permitted and the garden is to be used.
5. No talking during business hours.
6. The management expect a rise in output to compensate for these near Utopian conditions.

It is not recorded for how long these regulations lasted, or were obeyed, but the means of heating continued and one example is in a drawing office during the mid 20th century, where it was the apprentice's job, in addition to making the tea, to maintain the coal fire in the hearth.

Due to the arduous working conditions there was a great temptation to avoid going to work and in the Black Country this perspective was given a label. Monday was known as St. Monday due to absence as a result of recovering from the excesses of Sunday. Half of the employees do no work on Tuesdays. Wednesday is market day and an excuse to do half a day's work. As a consequence of attending the market many are unfit for work on a Thursday. Full work takes place on Fridays and Saturday mornings. Thankfully this pattern of 'anti-work' has passed away although, for many people, there remains the 'Monday morning feeling'.

8:3 Companies and Trade

The 19th century saw massive developments in technology in Great Britain evolving from the pioneering work beforehand. The opportunity was there uniquely in Britain, in contrast to other countries. China had become stagnated after its ancient superiority. Religious culture inhibited innovation. Europe fell behind due to lack of liberty, political interference or revolution. Against this background Britain was able to flourish and trade. The resources of skilled manpower and a supply of natural resources, together with the relative insularity of being on an island, conspired to generate wealth. In previous centuries Britain supplied raw materials to foreign artisans who then sold it back to England as finished products. They also gave us culture – pictures and music. The continent of Europe had a head start in science and engineering but this was about to change towards the end of the 18th century.

Machinery was introduced and some craftsmen saw this as a threat to their livelihood, especially the hand weavers, where prices would be undercut and inferior goods produced. Machines were destroyed and premises burnt. Resentment was felt by workers when mechanisation relieved them of menial tasks; they were unable to seize the opportunity that the machine provided to enable them to progress to more worthwhile and satisfying activities. Some craftsmen, on the other hand, recognised that the machine performed work quickly and still allowed the exercise of their traditional skills.

The transfer of skills tended to follow a logical pattern. Joiners were to become millwrights and turners; smiths became foundrymen; clockmakers were tool and die cutters. The towns became flooded with the mechanical organisation of industry, where work was readily available in order to earn a livelihood and gain money to improve the standard of living. If an employer required someone to work on a Sunday he would be paid double to compensate for depriving him from attending church and having recreation (re-creation). Working on a Sunday now is a way to increase earnings.

Where individuals had a mutual commitment, especially by sharing the views of a persecuted religion, they were seen to be as valuable in a company as a blood relative, provided that they possessed the necessary skills required by the company. The values of the non-conformist faith of diligence, thrift, and integrity were recognised.

Sheffield thrived on the production of cutlery where the craftsmen worked independently as sub-contractors to the large companies. They were vulnerable in times of economic depression and had to produce first rate quality for minimum cost. They dominated the British hand tool trade and their wares were exported to every part of the world. A gross (144) of knives and forks was sold to the Americans for less than three knives and forks could be bought at a retail store in America.

Geographical areas had particular requirements and local engineering companies satisfied those needs. Agricultural equipment was needed in East Anglia. Marine engine manufacture and shipbuilding took place on the Clyde, Tees and Mersey. Willenhall in Staffordshire became the centre of lock manufacture.

Sites adjacent to flowing water, iron ore and limestone deposits, with access to markets, were desirable for engineering works where products in iron and steel could be made entirely; from raw material to finish machined. Locally based skills were enriched by immigrants from Europe, the Huguenots and particularly the Flemish people.

Talent was evident in Manchester where many of the population were gifted with mechanical instinct. It was claimed that this pedigree was attributed to descent from the Norman smiths and armourers who occupied the area in the 11th century. The names of tools and implements could be traced to old Norman/French words. The climate in northern England was conducive to working cotton, being cool with high humidity. This had the effect of making the cotton pliable without losing its strength, so the industry prospered in that area.

Bridport in Dorset was the centre for producing rope, where up to 8,000 people were employed. The climate and soil was suitable for the growth of hemp and flax and the industry had flourished since the 13th century. With the advent of iron cable chains in the early 19th century, due to the Royal Navy requirements for a stronger means of anchoring and mooring ships, the manufacture moved to the Black Country, particularly Cradley Heath. An iron chain of 2 inches diameter was equivalent in strength to the previous hemp rope of 8 inches diameter.

Richard Roberts was a partner in Sharp and Roberts who made locomotive engines in Manchester and had at least 30 patents to his credit, gaining a reputation as an engineering genius of astonishing versatility. He made machine tools for the textile industry and is recognised as the founder of Production Engineering. Roberts' early employment was working at a lime quarry hauling wagons whilst practising woodwork in his spare time.

Josiah Wedgwood began his working life at the potter's wheel. He contracted smallpox, which led to complications, and he was laid up for many months. During this time he had the opportunity to think about his future. He was unable to continue his work as a potter after the illness and devoted his life to developing the world renowned company which bears his name. There are other examples where 'out of adversity comes good fortune and prosperity'.

Redditch in Worcestershire was the leading source of supply of needles for the hosiery and clothing industry. The source of power for needle making came from water mills converted from corn mills along the River Arrow valley. The craftsmen settled there after fleeing the restrictive trade guilds of London and did not wish to associate too closely with nearby Birmingham. A small needle making company operated at Hathersage in Derbyshire, owned by Henry Cocker who ran it under strict puritan values.

The partnership of Boulton and Watt developed into a highly successful company. It worked because Watt managed the technical side and Boulton attended to the company issues, and when joined by William Murdock and a skilled workforce they achieved remarkable results.

James Watt's mind was closed to new ideas (from others) and as previously mentioned, his steam engine patent was extended to the year 1800. Any enterprise by other engineers, particularly those in Cornwall, was prevented by action through the courts. Consequently development was stifled and technological progress delayed.

Matthew Murray was an inventive genius working in Leeds and he demanded exacting standards of craftsmanship, freely sharing the benefit of his experience at a time when frankness (as viewed by James Watt) was rare in business.

In 1784 William Pitt, prime minister, proposed a tax on the raw materials used in engineering. Matthew Boulton objected strongly to this, claiming that taxes should be levied on luxuries, riches and property; not on the means of getting them. He organised a petition with other influential people to prevent the imposition of this tax and also to stop export duties on manufactured goods. This was at a time when foreign governments were discouraging British manufacture by imposing heavy import duties.

Intervention by government in an industry which was gaining momentum, with prosperity growing, communications improving and the machine tool business about to erupt, must have been deeply distressing to the entrepreneurs. Any successful commercial activity can be a target to gain revenue provided that it is not detrimental to the long term development.

In the middle of the 18th century there was a considerable quantity of counterfeit money circulating in Birmingham. A significant trade had been established since the machinery used in the toy business, making buckles, buttons, snuff boxes, etc., could easily be adapted to stamp out coins by unscrupulous companies. The public were aware of this and could recognise a coin of 'Brummagem' origin and the authorities reacted by enforcing punishment ranging from imprisonment to hanging. Matthew Boulton, as owner of one of the largest and most renowned companies in Birmingham, refused all approaches to get involved in the disreputable business and eventually obtained a government contract to produce coinage which replaced the old issue and stopped counterfeit. The officials of the Mint took ten years to accept the idea of a private individual producing coins of the realm. Boulton also struck coinage for foreign governments and produced commemorative medals. In this work he was assisted in the artwork by sculptors and artists with technical experience supplied by an expert French die-sinker.

Pen making was carried out in Birmingham by Mitchell. Gillott came along and was given the details of the process by Mitchell's daughter and he then allegedly secretly developed pen making to a better quality and greater quantity than Mitchell. Gillott married Mitchell's daughter and on the morning of their wedding day, before going to the church, he apparently made a gross (144) of pens and sold them at one shilling each.

Isaac Wilkinson, the son of John Wilkinson the iron master, created an iron works at Merthyr Tydfil – the Dowlais Iron Company – and recommended John Guest as the company manager. John Guest subsequently created GKN (Guest, Keen and Nettlefold), a company which had a virtual monopoly in the manufacture of metal fasteners – screws, bolts, nuts etc. Henry Maudslay's reputation was such that he was regarded as the centre of all that was excellent in mechanical engineering. He was fond of maxims and would quote:

> 'Get a clear notion of what you desire to accomplish and then in all probability you will succeed in doing it.'
>
> 'Avoid complexities and make everything as simple as possible.'

The representation of machine tools and other products on paper, i.e. drawings, grew slowly through the 19th century. They were, in effect, pictures without dimensions or text. In the early days scale models were initially made and these were used to communicate the wishes of the customer to the workshop. Scale models have to give a realistic impression of the finished product and for this reason the degree of scale sometimes appears illogical. A five inch gauge model locomotive is 5:48 and ratios of 1:6 and 1:16 are also used and create an appearance of 'looking right'.

Joseph Whitworth did not commit records to paper and memorised data, ideas and designs. James Nasmyth, on the other hand, having inherited an ability in art from his father, was able to produce detailed drawings with good perspective representations of products. Mid 19th century drawings of machines were beautiful works of art. Pen and ink

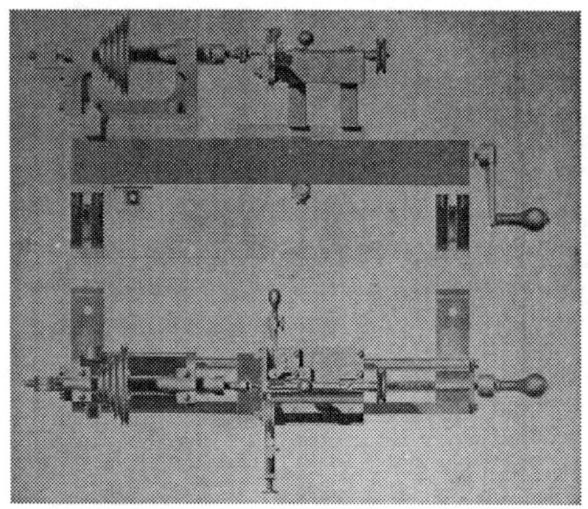

Pen and ink drawing with colour wash, 1845.

finely executed with colour washes graded to indicate curvature where necessary were (and still are) a joy to behold.

Trade associations were created to protect interests and to discuss and promote ideas. One of the most prominent clubs was the Lunar Society whose members were distinguished men including James Watt, Matthew Boulton, Josiah Wedgwood and Erasmus Darwin. They met for dinner at a member's house once a month on a Monday nearest to the time of the full moon, the reason being that they could take advantage of the moonlight to journey home after the meeting, hence the name of the society.

Craft societies were concerned in protecting their members against dilution and descent into dishonourable sectors of the trade, where skills were not valued and the market was flooded with men prepared to work long hours for low wages producing inferior goods. The foundation of the Mechanics Institutes during the 1820s to popularise scientific knowledge and to teach literacy appealed to artisans. The Anglican Church was initially opposed to such a venture as (they thought) it would encourage disaffection and breed freethinkers. The Institutes thrived and improved the skill and practice of those who had the potential to develop, and then by their industry to contribute to the prosperity of society in general. The artisans lived in the more prosperous suburbs of a town on the western fringe. The prevailing wind reached these areas first and they were spared the stench from industry and the hovels.

Despite the earlier inventions and innovations there were many small companies which thrived during the late 19th century using outmoded techniques. Little investment was made in new equipment; drawings were often inadequate; measuring equipment crude; tooling 'cobbled' to suit; lighting by gas lamp poor and coolant on the lathe was dispensed by dripping onto the tool and supplemented occasionally by accurate squirts of tobacco juice. Surprisingly, excellent work was produced.

Finally, this account has been an attempt to emphasise, by just a few examples, the degree of engineering craftsmanship achieved in the past and the evidence shown deserves our greatest admiration.

Conclusion

This narrative has been a brief glimpse into the Industrial Revolution period with eclectic examples of an interesting and unusual nature. Although the main theme was an appreciation of engineering craftsmanship, the references to other activities at another time, perhaps not exactly relevant, can be justified for inclusion due to their curious and peculiar nature and their evidence of expertise. As children we are all endowed, to a greater or lesser extent, with an inquisitive nature and with maturity this curiosity can be developed to see beyond the superficial. That which at first glance appears insignificant can often prove to be fascinating and worthy of our discerning attention. There is lots of scope for the reader to delve more deeply into the subject matter, and hopefully this will now have been encouraged so that a greater sharing of the excitement of the past may be gained by a wider audience.

The engineers, who evolved from craftsmen, had a difficult time in getting acceptance of their proposals for change. They had foresight but often had to wait for a more propitious time for implementation, sometimes posthumously.

History is monopolised by the deeds of monarchs and statesmen and their memorials abound. The pioneering engineers have few monuments by which succeeding generations may be reminded of their contribution to our prosperity. Christopher Wren is buried in St. Paul's Cathedral, London and above his tomb is inscribed, 'Reader, if you seek his monument – look around you.'

We are fortunate to have retained some physical evidence of past work and can often, by conjecture or surmise, deduce the ‘what?’, ‘why?’ and ‘how?’ aspects of an artefact, always provided that the depredations of time and the elements have not had their influence. Our gratitude must be accorded to those individuals and societies who are dedicated to the preservation and recording of the past to ensure appreciation and understanding in the future.

In addition to the acquisition of skill and the dedication to achieve perfection, another feature contributes to the craftsman’s work and that is a spiritual dimension. There is a sense of the ethereal about a piece of work which has been commissioned by a particularly special client, which is not entirely as evident as the artistry and quality of the work. This rather uncanny aspect may be partly explained by the fact that, when working on a piece of work, the craftsman is conscious of the nature of the client and somehow the essence of this pervades the work.

Finally, a quotation attributed to St. Francis of Assisi:

> ‘He who works with his hands is a labourer.
> He who works with his hands and his head is a craftsman.
> He who works with his hands and his head and his heart is an artist.’

There are examples in this account where the craftsman may justifiably be described as an artist.

Index

Bibliography

T. W. Barber – *The Engineer's Sketchbook of Mechanical Movements*, 1889

Rev. F. Barlow – *The Complete English Dictionary*, 1772

Bible – *The Apocrypha*; Sirach, 38:24–34

R. Buchanan – *Practical Essays on Mill Work*, 1823

James Nasmyth – *Autobiography*, 1883

Samuel Smiles – *Invention and Industry*, 1884

Samuel Smiles – *Lives of the Engineers*, 1865